人工智能开发与实战丛书

Python 数据分析入门

［英］何塞·安平科（José Unpingco） 著

安 翔 刘 强 译

机械工业出版社

本书系统性地总结了 Python 在数据分析中的应用方法，围绕数据分析所需的编程技能进行了详细的讲解与实践指导。本书从基本编程出发，逐步深入到面向对象编程、使用模块、数组操作、数据处理以及可视化数据等关键领域，并通过丰富的示例展示了如何将 Python 的核心工具高效地应用于复杂的数据分析场景。本书注重实践与理论结合，帮助读者建立对数据分析任务的系统性理解。

本书面向具有一定 Python 基础的读者，包括数据分析、机器学习及相关领域的研究人员、工程师以及高等院校的高年级本科生和研究生。此外，本书也适合希望通过掌握数据分析工具提升项目实践能力的读者阅读。无论是希望夯实编程基础的读者，还是希望在数据分析领域深入探索的技术人员，都可以通过本书获得清晰的思路和实用的工具支持。

First published in English under the title
Python Programming for Data Analysis
by José Unpingco, edition: 1
Copyright © José Unpingco
This edition has been translated and published under licence from
Springer Nature Switzerland AG.

此版本仅限在中国大陆地区（不包括香港、澳门特别行政区及台湾地区）销售。未经出版者书面许可，不得以任何方式抄袭、复制或节录本书中的任何部分。

北京市版权局著作权合同登记　图字：01-2022-1577 号。

图书在版编目（CIP）数据

Python数据分析入门 /（英）何塞·安平科著 ；安翔，刘强译. -- 北京 ：机械工业出版社，2025.4.
(人工智能开发与实战丛书). -- ISBN 978-7-111-78202-5

I. TP312.8

中国国家版本馆CIP数据核字第20251RZ682号

机械工业出版社（北京市百万庄大街22号　邮政编码100037）
策划编辑：付承桂　　　　　　责任编辑：付承桂　杨　琼
责任校对：曹若菲　陈　越　　封面设计：马精明
责任印制：单爱军
北京华宇信诺印刷有限公司印刷
2025年6月第1版第1次印刷
169mm×239mm・16印张・272千字
标准书号：ISBN 978-7-111-78202-5
定价：99.00元

电话服务　　　　　　　　　网络服务
客服电话：010-88361066　　机 工 官 网：www.cmpbook.com
　　　　　010-88379833　　机 工 官 博：weibo.com/cmp1952
　　　　　010-68326294　　金　书　网：www.golden-book.com
封底无防伪标均为盗版　　　机工教育服务网：www.cmpedu.com

译 者 序

本书是一本深入探讨 Python 在数据分析中应用的精彩之作。通过对基本编程、面向对象编程、使用模块、Numpy、Pandas 和可视化数据内容的全面介绍，本书将为那些希望在数据分析领域发展的读者提供有益的指导和实用的技能。通过逐步引导和详细讲解，本书将帮助读者建立坚实的编程基础，掌握 Python 等编程语言的强大功能。

本书第 1 章从基础语言入手，介绍了编程中的关键概念和基本数据结构，涵盖了循环和条件、函数、文件输入 / 输出等内容，同时深入讨论了 Python 的高级功能，如生成器、装饰器、迭代等，帮助读者打下坚实的编程基础。第 2 章探讨了属性、方法、继承等面向对象编程的重要概念，帮助读者理解并运用面向对象编程的思想来解决问题，同时介绍了元编程、设计模式等高级主题。第 3 章详细介绍了如何使用标准库以及编写和使用自己的模块，并介绍了动态导入、从 Web 中获取模块等内容，帮助读者更好地组织和管理自己的代码。第 4~6 章分别介绍了 Numpy、Pandas 和可视化数据的相关知识，涵盖了数据处理、数据分析和数据可视化等方面，帮助读者更好地处理和展示数据，从而更深入地理解数据背后的含义。

通过本书的学习，读者将掌握编程的基础知识和技能，能够利用编程解决实际问题，进行数据处理和分析，并将数据可视化呈现，为读者打开编程世界的大门，让他们能够更好地应对未来的挑战和机遇。祝愿读者在本书中找到所需的知识和获得启发，展开编程之旅，探索无限可能。

本书由安翔负责第 1~6 章的翻译工作，刘强负责前言内容的翻译工作以及全书的统稿工作。同时，特别感谢研究生薄浩、魏叶茹以及贺忱为本书的校对工作所做出的贡献。

本书的出版得到了北京市教育委员会项目（22019821001）、北京石油化工学院致远科研基金项目（No.2024104）的资助，在此表示衷心的感谢。最后，真诚感谢付承桂、杨琼编辑为本书出版所付出的努力。

由于译者水平有限，若有不妥之处，敬请批评指正。

前　言

本书是基于我在加州大学圣地亚哥分校教授的 ECE143 数据分析编程课程的讲义而撰写的，该课程是机器学习和数据科学研究生和本科生的必修课程。本书专门针对在数据分析领域中 Python 的用语和方法进行讲解，适用于有一定编程基础和编程经验（如 MATLAB 或 Java）的读者。具体而言，本书重点介绍如何使用特定的 Python 语言来对混乱且复杂的原始数据进行整理，以及如何高效地完成数据预处理工作。

在数据分析编程课程的教学过程中，我发现向学生进行相关知识点的背景讲解和讨论非常有助于初学者在代码工程中做出更好的选择。因此，本书除了讲解如何使用 Python 进行数据分析以外，还讲解为什么这样进行数据分析。另外，本书还提供了一些小技巧，帮助读者创作出适合在实际生产和开发中使用的具有可读性和可维护性的代码。

本书重点介绍了如何有效地使用 Python 语言，然后介绍了一些关键的第三方模块。首先，本书对 Numpy 数值数组模块进行了详细讲解，因为它是 Python 中所有数据科学和机器学习的基础。接下来，本书介绍了 Pandas 模块，使用其功能来实现有效和流畅的数据处理。由于数据可视化对于数据科学和机器学习至关重要，因此本书还详细介绍了 Matplotlib 模块以及基于 Web 的 Bokeh、Holoviews、Plotly 和 Altair 等第三方模块。

本书非常适合已经具备一些 Python 基础，并想通过了解 Python 的工作方式和原因来提高自身水平的读者。要想充分利用本书，请打开 Python 解释器并开始键入本书提供的代码示例。

致 谢

我想感谢 Jupyter Notebook 的两位创始人 Brian Granger 和 Fernando Perez，感谢他们所做出的出色工作，同时还想感谢整个 Python 社区，正是他们的无私贡献使得本书成为可能。本书使用了 Doconce[1] 文档准备系统，其作者是 Hans Petter Langtangen。感谢 Geoffrey Poore[2] 在 PythonTeX 和 LATEX 方面的工作，这些都是本书编写中所用到的关键技术。

José Unpingco

美国加利福尼亚州圣地亚哥

2020 年 2 月

参考文献

1. H.P. Langtangen, DocOnce markup language. https://github.com/hplgit/doconce
2. G.M. Poore, Pythontex: reproducible documents with latex, python, and more. Comput. Sci. Discov. **8**(1), 014010 (2015)

目　录

译者序

前言

致谢

第 1 章　基本编程 ··· 1

　　1.1　基础语言 ··· 1
　　　　1.1.1　入门 ·· 2
　　　　1.1.2　保留关键字 ·· 3
　　　　1.1.3　数字 ·· 3
　　　　1.1.4　复数 ·· 4
　　　　1.1.5　字符串 ··· 5
　　　　1.1.6　基本数据结构 ·· 9
　　　　1.1.7　循环和条件 ·· 16
　　　　1.1.8　函数 ·· 20
　　　　1.1.9　文件输入/输出 ·· 32
　　　　1.1.10　处理错误 ··· 34
　　　　1.1.11　掌握 Python 的强大功能 ·· 37
　　　　1.1.12　生成器 ··· 40
　　　　1.1.13　装饰器 ··· 45
　　　　1.1.14　迭代 ·· 49
　　　　1.1.15　使用 Python 断言进行预调试 ··· 57
　　　　1.1.16　使用 sys.settrace 进行堆栈追踪 ·· 58
　　　　1.1.17　使用 IPython 进行调试 ··· 59
　　　　1.1.18　从 Python 中进行日志记录 ··· 59

第 2 章　面向对象编程 ··· 62

- 2.1　属性 ··· 62
- 2.2　方法 ··· 64
- 2.3　继承 ··· 65
- 2.4　类变量 ··· 67
- 2.5　类函数 ··· 68
- 2.6　静态方法 ··· 70
- 2.7　哈希对子变量隐藏父变量 ··· 70
- 2.8　委托函数 ··· 71
- 2.9　使用 super 进行委托 ··· 71
- 2.10　元编程：猴子补丁 ··· 73
- 2.11　抽象基类 ··· 74
- 2.12　描述符 ··· 76
- 2.13　具名元组和数据类 ··· 79
- 2.14　泛型函数 ··· 82
- 2.15　设计模式 ··· 84
  - 2.15.1　模板 ··· 85
  - 2.15.2　单列模式 ··· 85
  - 2.15.3　观察者 ··· 86
  - 2.15.4　适配器 ··· 87
- 参考文献 ··· 87

第 3 章　使用模块 ··· 88

- 3.1　标准库 ··· 88
- 3.2　编写和使用自己的模块 ··· 90
  - 3.2.1　将目录用作模块 ··· 91
- 3.3　动态导入 ··· 91

3.4 从 Web 中获取模块 ··· 92

3.5 Conda 包管理 ··· 92

参考文献 ··· 94

第 4 章 Numpy ··· 95

4.1 Dtypes ··· 95

4.2 多维数组 ··· 96

4.3 重塑和堆叠 Numpy 数组 ··· 97

4.4 复制 Numpy 数组 ··· 98

4.5 切片、逻辑数组操作 ··· 99

4.6 Numpy 数组和内存 ··· 100

4.7 Numpy 内存数据结构 ··· 103

4.8 数组元素操作 ··· 105

4.9 通用函数 ··· 106

4.10 Numpy 数据输入 / 输出 ··· 107

4.11 线性代数 ··· 107

4.12 广播 ··· 108

4.13 掩码数组 ··· 112

4.14 浮点数 ··· 113

4.15 高级 Numpy dtypes ··· 116

参考文献 ··· 117

第 5 章 Pandas ··· 118

5.1 使用 Series ··· 118

5.2 使用数据帧 ··· 121

5.3 重新索引 ··· 125

5.4 删除项目 ··· 127

- 5.5 高级索引 ········ 127
- 5.6 广播和数据对齐 ········ 128
- 5.7 分类和合并 ········ 131
- 5.8 内存使用和数据类型 dtypes ········ 133
- 5.9 常见的操作 ········ 136
- 5.10 显示 DataFrame ········ 137
- 5.11 分层索引 ········ 139
- 5.12 Pipes ········ 142
- 5.13 数据文件和数据库 ········ 142
- 5.14 自定义 Pandas ········ 143
- 5.15 滚动和填充操作 ········ 144

第 6 章 可视化数据 ········ 146

- 6.1 Matplotlib ········ 147
  - 6.1.1 设置默认值 ········ 149
  - 6.1.2 图例 ········ 149
  - 6.1.3 子图 ········ 149
  - 6.1.4 Spines ········ 150
  - 6.1.5 共享轴 ········ 151
  - 6.1.6 三维曲面 ········ 152
  - 6.1.7 使用 patch ········ 153
  - 6.1.8 3d 中的 patches ········ 153
  - 6.1.9 使用 transformation ········ 155
  - 6.1.10 使用文本注释 ········ 158
  - 6.1.11 使用箭头注释 ········ 158
  - 6.1.12 嵌入可缩放/不可缩放的子图 ········ 161
  - 6.1.13 动画 ········ 163
  - 6.1.14 直接使用路径 ········ 164
  - 6.1.15 使用滑块与绘图交互 ········ 167
  - 6.1.16 色彩图 ········ 168
  - 6.1.17 使用 setp 和 getp ········ 169

目 录

- 6.1.18 与 Matplotlib 图形交互 ······ 170
- 6.1.19 键盘事件 ······ 170
- 6.1.20 鼠标事件 ······ 172

6.2 Seaborn ······ 173
- 6.2.1 自动聚合 ······ 176
- 6.2.2 多个绘图 ······ 180
- 6.2.3 分布图 ······ 181

6.3 Bokeh ······ 190
- 6.3.1 使用 Bokeh 基元 ······ 190
- 6.3.2 Bokeh 布局 ······ 192
- 6.3.3 Bokeh 组件 ······ 194

6.4 Altair ······ 199
- 6.4.1 Altair 细节化 ······ 201
- 6.4.2 聚合和转换 ······ 203
- 6.4.3 Altair 交互 ······ 207

6.5 Holoviews ······ 210
- 6.5.1 数据集 ······ 214
- 6.5.2 图像数据 ······ 216
- 6.5.3 表格数据 ······ 217
- 6.5.4 自定义交互 ······ 219
- 6.5.5 流 ······ 220
- 6.5.6 Pandas 与 hvplot 集成 ······ 221
- 6.5.7 网络图 ······ 226
- 6.5.8 Holoviz Panel ······ 231

6.6 Plotly ······ 234

参考文献 ······ 242

第 1 章

基本编程

1.1 基础语言

在深入了解 Python 之前，对 Python 语言有个整体性的了解将有助于你以后在软件开发过程中做出更好的决策，特别是当你的项目变得越来越庞大和复杂时。Python 起源于 20 世纪 80 年代在荷兰开发的一种名为 ABC 的语言，该语言作为 BASIC 的代替品，旨在让非计算机专业的科学家们能更有效地使用当时新兴的微型计算机。事实上，Python 作为 ABC 语言的派生物，其实用主义精神一直贯穿于其演变中。

"因语言而来，为社区而留"——Python 是一个开源项目，由社区驱动，因此并没有实体公司为该语言的发展做从上到下的决策。这种发展方式原本应该会导致混乱，但其创始人 Guido van Rossum 的耐心和务实的领导使 Python 不断受益。在 Guido 退休后，他的领导角色由一个独立的治理委员会来接替。多年来，Python 语言因其开放性设计和高质量的源代码吸引了世界各地才华横溢的开发者为丰富其标准库做出贡献。同时，Python 也以其社区对新手的友好而闻名，新手很容易在网上找到关于 Python 入门的帮助。

语言的务实性和强大的社区支持使 Python 成为开发 Web 应用程序的一种非常好的方案。在数据科学和机器学习出现之前，Python 社区中高达 80% 的参与者是 Web 开发者，因此 Python 标准库中有许多 Web 协议和技术。但在过去五年中（在本书撰写时），Python 社区中 Web 开发者的数量和数据科学家的数量正逐渐持平。

  Python 是一种解释型语言，与编译型语言（如 C 或 FORTRAN）不同。虽然这两种类型的语言都需要源代码文件，但编译型语言需要编译器来全面检查源代码并生成一个链接到系统特定库文件的可执行文件。一旦可执行文件创建完成，就不再需要编译器，只需在系统上运行可执行文件即可。而 Python 这样的解释型语言，则必须始终运行 Python 进程才能执行代码。这是因为 Python 进程对其运行的平台而言是抽象的，必须对源代码中的指令进行解释后才能在平台上执行它们。而 Python 解释器作为源代码和平台之间的中介，负责承担这个解释任务。只要每个平台上都有 Python 解释器，那么源代码就可以在不同的平台上运行，这样 Python 源代码在平台之间就具有了可移植性，而这正是解释型语言的关键优势。而编译型语言由于其可执行文件与平台及特定库文件相关，所以其代码的可移植性比 Python 要弱。但编译型语言的优势是编译器能够有选择地利用平台，并可以对源代码文件进行学习和优化，从而加速所生成的可执行文件。总的来说，这就是解释型语言和编译型语言之间的主要区别。

  有些人认为 Python 比编译型语言效率低，这要取决于如何来理解效率问题。因为如果从开发时间而非仅是代码运行时间来计算，Python 显然更快一些，它的开发迭代周期不再需要繁琐的编译和链接步骤。此外，Python 比编译型语言更容易使用，因为许多棘手的问题（如内存管理）都被自动处理了。Python 的快速迭代对于追求效率的产品开发来说是一个重要优势。而且，Python 不太适合对算力敏感的应用场景，例如解决并行微分方程组、模拟大规模流体力学或其他大规模物理计算等，但 Python 可在这些应用场景中用来做数据的后处理工作。

1.1.1 入门

  Python 解释器的主要接口是命令行。你可以直接在终端中输入 Python，并将会看到类似以下内容的代码：

```
Python 3.7.3 (default, Mar 27 2019, 22:11:17)
[GCC 7.3.0] :: Anaconda, Inc. on linux
Type "help", "copyright", "credits" or "license" for more
↪   information.
>>>
```

上面的代码提供了许多有用的信息，包括 Python 的版本以及其来源。这很重要，因为有时 Python 解释器会允许快速访问某些特定模块（例如 math 模块）。

1.1.2 保留关键字

在对变量或函数命名时，请避免使用以下的保留关键字：

```
and        del        from       not        while
as         elif       global     or         with
assert     else       if         pass       yield
break      except     import     print
class      exec       in         raise
continue   finally    is         return
def        for        lambda     try
```

同时还有这些：

```
abs all any ascii bin bool breakpoint bytearray bytes callable
chr classmethod compile complex copyright credits delattr
dict dir display divmod enumerate eval exec filter float
format frozenset getattr globals hasattr hash help hex id
input int isinstance issubclass iter len list locals map max
memoryview min next object oct open ord pow print property
range repr reversed round set setattr slice sorted
staticmethod str sum super tuple type vars zip
```

例如，一个常见的错误是分配 sum=10，这会导致 Python 中的 sum（ ）函数不再可用。

1.1.3 数字

Python 具有简单的数字处理能力。在 Python 中，井号（#）表示注释。

```
>>> 2+2
4
>>> 2+2    # 注释应与代码在同一行
4
>>> (50-5*6)/4
5.0
```

注意，Python 2 中的除法是整数除法，而 Python 3 中是浮点型除法，如需整数除法，可以使用 // 代替 /。Python 是动态类型语言，因此可以像下面这样对 width 和 height 直接赋值，而无需提前声明 width 和 height 的类型。

```
>>> width = 20
>>> height = 5*9
>>> width * height
900
>>> x = y = z = 0    # 将 x、y 和 z 赋值为零
```

```
>>> x
0
>>> y
0
>>> z
0
>>> 3 * 3.75 / 1.5
7.5
>>> 7.0 / 2 # 浮点计算
3.5
>>> 7/2
3.5
>>> 7 // 2 # 双斜杠实现整数除法
3
```

我们可以将这种动态赋值方式理解为对内存中的值赋予一个标签。例如，width 是数字 20 的标签。Python 3.8 引入了一种叫海象运算符（walrus）的新赋值运算符（:=），允许赋值本身具有赋值对象的值，如下所示：

```
>>> print(x:=10)
10
>>> print(x)
10
```

该运算符还有许多其他微妙的用法，可在某些情况下提高代码的可读性。Python 中还可以对数值类型进行直接的转换，如下所示：

```
>>> int(1.33333)
1
>>> float(1)
1.0
>>> type(1)
<class 'int'>
>>> type(float(1))
<class 'float'>
```

值得注意的是，Python 中的整数长度是没有限制的。这是因为它们在内部存储为数字列表，所以它们的操作速度比具有固定位长度的 Numpy 整数慢。

> **编程技巧：IPython**
>
> IPython 解释器提供了诸如制表符补全等非常方便的功能，可以让你的 Python 编程更加容易。
>
> 请参见 http://github.com/ipython/ipython 获取如何开始使用 IPython 的最新信息。

1.1.4 复数

Python 可支持基本复数运算。

```
>>> 1j * 1J
(-1+0j)
>>> 1j * complex(0,1)
(-1+0j)
>>> 3+1j*3
(3+3j)
>>> (3+1j)*3
(9+3j)
>>> (1+2j)/(1+1j)
(1.5+0.5j)
>>> a=1.5+0.5j
>>> a.real  # 句点"·"用来访问对象的属性
1.5
>>> a.imag
0.5
>>> a=3.0+4.0j
>>> float(a)
Traceback (most recent call last):
  File "<stdin>", line 1, in <module>
TypeError: can't convert complex to float
>>> a.real
3.0
>>> a.imag
4.0
>>> abs(a)    # sqrt(a.real**2 + a.imag**2)
5.0
>>> tax = 12.5 / 100
>>> price = 100.50
>>> price * tax
12.5625
>>> price + _
113.0625
>>> # 下划线字符"_"指代最近一次计算的结果值
>>> round(_, 2) # 下划线字符"_"指代最近一次计算的结果值
↪   result
113.06
```

通常情况下，我们建议使用 Numpy 进行复数计算。

1.1.5 字符串

Python 对字符串的处理进行了高度的优化，在这里只简单介绍一些要点。首先，单引号或双引号都可以定义一个字符串，它们之间不存在优先级关系。

```
>>> 'spam eggs'
'spam eggs'
>>> 'doesn\'t'  # 反斜杠作为转义字符
"doesn't"
>>> "doesn't"
"doesn't"
>>> '"Yes," he said.'
'"Yes," he said.'
>>> "\"Yes,\" he said."
'"Yes," he said.'
>>> '"Isn\'t," she said.'
'"Isn\'t," she said.'
```

Python 中对字符串的操作延续了 C 语言风格的换行符、制表符等转义字符。Python3 中字符串默认使用 UTF-8 编码，而不是 Python 2 中的 ASCII 编码。在对函数进行文档说明时，使用文档字符串（docstrings）。文档字符串可以用三个单引号（'''）或三个双引号（"""）来申明。

```
>>> print( '''Usage: thingy [OPTIONS]
... and more lines
... here and
... here
...        ''')
Usage: thingy [OPTIONS]
and more lines
here and
here
```

Python 对字符串的操作可以通过在单引号或双引号之前加上特定字符来进行控制。例如下面的代码，注释（#）中对此操作进行了说明。

```
>>> # 出现在双引号前的'r'使下面的字符串成为一个"原始(raw)"字符串
>>> hello = r"This long string contains newline characters \n, as
↪ in C"
>>> print(hello)
This long string contains newline characters \n, as in C
>>> # 如果没有'r',你会得到一个出现换行的字符串
>>> hello = "This long string contains newline characters \n, as
↪ in C"
>>> print(hello)
This long string contains newline characters
, as in C
>>> u'this a unicode string  μ ±'  # 'u'让 Python2 中的字符串采用 unicode 编码
'this a unicode string  μ ±'
>>> 'this a unicode string  μ ±'   # 在 Python3 中,不使用'u'也是默认 unicode 编码
'this a unicode string  μ ±'
>>> u'this a unicode string \xb5 \xb1' # 使用十六进制代码表示 μ ±
'this a unicode string μ ±'
```

Python 3 中还有一个非常有趣的字符串操作：f-string，它可以对 {} 中包含的变量进行替换。

```
>>> x = 10
>>> s = f'{x}'
>>> type(s)
<class 'str'>
>>> s
'10'
```

在字符串操作中，还有非常重要的一点，即 Python 字符串具有不可变性。这意味着字符串创建后无法再对其进行修改。示例如下：

```
>>> x = 'strings are immutable '
>>> x[0] = 'S' # 不允许对字符串对象进行修改
Traceback (most recent call last):
  File "<stdin>", line 1, in <module>
TypeError: 'str' object does not support item assignment
```

字符串与字节

在 Python 3 中，字符串的默认编码格式为 UTF-8。

注意：Python 3 中字节和字符串是不同的对象。示例如下：

```
>>> x='Ø'
>>> isinstance(x,str)      # 用 isinstance()函数来判断 x 是否是字符串
True
>>> isinstance(x,bytes)    # isinstance()函数来判断 x 是否是字节
False
>>> x.encode('utf8')       # 转换为字节
b'\xc3\x98'
```

我们也可以使用 decode() 将字节转换为字符串，示例如下：

```
>>> x=b'\xc3\x98'
>>> isinstance(x,bytes)    # 是
True
>>> isinstance(x,str)      # 否
False
>>> x.decode('utf8')
'Ø'
```

注意：在 Python 3 中，不支持类似 u "hello"＋b "goodbye" 这种字符串和字节的操作。如果需要进行这种操作，则必须先进行解码/编码，示例如下：

```
>>> x=b'\xc3\x98'
>>> isinstance(x,bytes)    # 是
True
>>> y='banana'
>>> isinstance(y,str)      # 否
True
>>> x+y.encode()
b'\xc3\x98banana'
>>> x.decode()+y
'Øbanana'
```

字符串切片

Python 是一种从零开始索引（与 C 类似）的语言。字符串切片可以使用冒号（:）字符实现，示例如下：

```
>>> word = 'Help' + 'A'
>>> word
'HelpA'
>>> '<' + word*5 + '>'
'<HelpAHelpAHelpAHelpAHelpA>'
>>> word[4]
'A'
>>> word[0:2]
'He'
>>> word[2:4]
'lp'
>>> word[-1]      # 最后一个字符
'A'
>>> word[-2]      # 倒数第二个字符
'p'
>>> word[-2:]     # 最后两个字符
```

```
'pA'
>>> word[:-2]      # 最后两个字符以外的字符
'Hel'
```

字符串运算

一些基本的数值运算也可以用来处理字符串。

```
>>> 'hey '+'you'   # 用加法运算符来实现两个字符串的连接
'hey you'
>>> 'hey '*3       # 用整数乘法运算符来实现复制字符串
'hey hey hey '
>>> ('hey ' 'you') # 使用无分隔逗号的小括号来实现两个字符串的连接
'hey you'
```

Python 拥有一个内置的、非常强大的正则表达式模块（re）用于字符串操作。字符串替换会创建新的字符串。

```
>>> x = 'This is a string'
>>> x.replace('string','newstring')
'This is a newstring'
>>> x # x 不变
'This is a string'
```

格式化字符串

在 Python 中有很多格式化字符串的方式。下面代码中是最简单的一种方法，它遵循 C 语言 sprintf 的约定，并结合模运算符 %：

```
>>> 'this is a decimal number %d'%(10)
'this is a decimal number 10'
>>> 'this is a float %3.2f'%(10.33)
'this is a float 10.33'
>>> x = 1.03
>>> 'this is a variable %e' % (x)  # 指数
'this is a variable 1.030000e+00'
```

可以用"+"直接连接字符串，示例如下：

```
>>> x = 10
>>> 'The value of x = '+str(x)
'The value of x = 10'
```

还可以用字典进行格式化，示例如下：

```
>>> data = {'x': 10, 'y':20.3}
>>> 'The value of x = %(x)d and y = %(y)f'%(data)
'The value of x = 10 and y = 20.300000'
```

用 format 函数进行格式化，示例如下：

```
>>> x = 10
>>> y = 20
>>> 'x = {0}, y = {1}'.format(x,y)
'x = 10, y = 20'
```

format 函数的优点是可以像下面这样重用占位符：

```
>>> 'x = {0},{1},{0};  y = {1}'.format(x,y)
'x = 10,20,10;  y = 20'
```

还有之前讨论过的 f-string。

> **编程技巧：Python 2 中的字符串**
>
> 在 Python 2 中，默认的字符串编码是 7 位 ASCII 编码。在字节和字符串之间没有区别。例如，你可以按照以下方式从二进制编码的 JPG 文件中读取内容：
> ```
> with open('hour_1a.jpg','r') as f:
>     x = f.read()
> ```
> 这在 Python 2 中可以正常工作，但在 Python 3 中会引发 UnicodeDecode-Error 错误。要在 Python 3 中解决这个问题，你需要使用 rb 二进制模式读取，而不是仅使用 r 文件模式。

1.1.6 基本数据结构

Python 提供了很多强大的数据结构，其中最强大的和最基础的是列表和字典。数据结构和算法是相辅相成的，如果你不了解数据结构，那么就不能有效地编写算法，反之亦然。从根本上讲，数据结构为程序员提供了一些保证，如果按照约定的方式使用数据结构，这些保证将得到满足。这些保证被称为数据结构的不变性。

列表

列表（list）是一种顺序保持型通用容器，用于实现序列数据结构。在索引一个非空列表时，列表总是会按顺序给出下一个有效元素，这是列表可靠性的体现。事实上，列表是 Python 的主要有序数据结构。这意味着，如果你遇到的问题中顺序很重要，那么就应该考虑使用列表数据结构。下面的示例将说明这一点。

```
>>> mix = [3,'tree',5.678,[8,4,2]]   # 可以包含子列表
>>> mix
[3, 'tree', 5.678, [8, 4, 2]]
>>> mix[0]      # 从 0 开始索引
3
>>> mix[1]      # 索引单个元素
'tree'
>>> mix[-2]     # 从右向左索引,倒数第 2 个
5.678
>>> mix[3]      # 子列表
[8, 4, 2]
>>> mix[3][1]   # 子列表中索引
4
```

```
>>> mix[0] = 666  # 列表是可变的
>>> mix
[666, 'tree', 5.678, [8, 4, 2]]
>>> submix = mix[0:3]  # 创建子列表
>>> submix
[666, 'tree', 5.678]
>>> switch = mix[3] + submix  # 用加法运算符添加列表
>>> switch
[8, 4, 2, 666, 'tree', 5.678]
>>> len(switch)   # 列表长度是内置函数
6
>>> resize=[6.45,'SOFIA',3,8.2E6,15,14]
>>> len(resize)
6
>>> resize[1:4] = [55]    # 更新列表中的部分元素
>>> resize
[6.45, 55, 15, 14]
>>> len(resize)  # 减少子列表
4
>>> resize[3]=['all','for','one']
>>> resize
[6.45, 55, 15, ['all', 'for', 'one']]
>>> len(resize)
4
>>> resize[4]=2.87   # 不能这样添加列表元素
Traceback (most recent call last):
  File "<stdin>", line 1, in <module>
IndexError: list assignment index out of range
>>> temp = resize[:3]
>>> resize = resize + [2.87] # 添加列表元素
>>> resize
[6.45, 55, 15, ['all', 'for', 'one'], 2.87]
>>> len(resize)
5
>>> del resize[3]      # 删除列表元素
>>> resize
[6.45, 55, 15, 2.87]
>>> len(resize)         # 列表变短
4
>>> del resize[1:3]    # 删除一个子列表
>>> resize
[6.45, 2.87]
>>> len(resize)         # 列表变短
2
```

> **编程技巧：列表排序**
>
>   Python 3 中内置的函数 sorted 可以对列表进行排序。
> ```
> >>> sorted([1,9,8,2])
> [1, 2, 8, 9]
> ```
>   也可以使用列表的 sort() 方法对列表进行排序。
> ```
> >>> x = [1,9,8,2]
> >>> x.sort()
> >>> x
> [1, 2, 8, 9]
> ```
>   这两种方法都使用了强大的 Timsort 算法。

在大概了解了如何索引和使用列表后，我们需要注意：在索引一个非空列表时，列表总是会按顺序给出下一个有效元素。例如：

```
>>> x = ['a',10,'c']
>>> x[1]  # 返回10
10
>>> x.remove(10)
>>> x[1]  # 下一个元素
'c'
```

注意，上面的代码中列表数据结构主动填补了移除10后产生的空白。此外，列表元素可通过整数索引访问，而整数具有自然顺序，因此列表也具有自然顺序。然而，维护列表的顺序索引并非没有代价，考虑以下情况：

```
>>> x = [1,3,'a']
>>> x.insert(0,10)  # 插入最前
>>> x
[10, 1, 3, 'a']
```

上面的代码是否看起来无害？对于小列表来说可能确实如此，但对于大列表来说则不然。这是因为要保持可靠性，列表必须将剩余的元素向右移动（即进行内存复制），以容纳在开头添加的新元素。在具有数百万个元素并且在循环中的大型列表上，这可能需要相当长的时间。这就是为什么默认的append()和pop()列表方法是对列表的末尾进行操作，因为在那里不需要将元素向右移动。

元组

元组是Python中另一个通用的序列容器，与列表非常相似，但它是不可变的。元组由逗号分隔（括号是分组符号），以下是一些示例：

```
>>> a = 1,2,3  # 无需括号
>>> type(a)
<class 'tuple'>
>>> pets=('dog','cat','bird')
>>> pets[0]
'dog'
>>> pets + pets  # 相加
('dog', 'cat', 'bird', 'dog', 'cat', 'bird')
>>> pets*3
('dog', 'cat', 'bird', 'dog', 'cat', 'bird', 'dog', 'cat', 'bird')
>>> pets[0]='rat'  # 无法修改
Traceback (most recent call last):
  File "<stdin>", line 1, in <module>
TypeError: 'tuple' object does not support item assignment
```

从上面代码可以看出，元组最重要的特点是它是不可变的，这使它对Python内存管理的开销更小。从这个意义上讲，元组更轻量级并更能为代码提供稳定性，这是元组的主要优势。

> **编程技巧：理解列表内存**
>
> Python 的 id 函数会显示一个与给定变量的内部引用相对应的整数。我们建议将变量赋值视为标签，因为在 Python 内部使用的是变量的 id，而不是变量名/标签。
>
> ```
> >>> x = y = z = 10.1100
> >>> id(x)    # 不同的标签,相同的 id
> 140271927806352
> >>> id(y)
> 140271927806352
> >>> id(z)
> 140271927806352
> ```
>
> 这对于可变的数据结构（如列表）更重要：
>
> ```
> >>> x = y = [1,3,4]
> >>> x[0] = 'a'
> >>> x
> ['a', 3, 4]
> >>> y
> ['a', 3, 4]
> >>> id(x),id(y)
> (140271930505344, 140271930505344)
> ```
>
> 因为 x 和 y 只是指向相同列表的两个标签，对一个标签所做的更改会影响另外一个。Python 在分配新内存方面本质上是吝啬的，因此如果你想要具有相同内容的两个不同列表，可以通过以下方式强制进行复制：
>
> ```
> >>> x = [1,3,4]
> >>> y = x[:]       # 强制复制
> >>> id(x),id(y) # 不同的 id
> (140271930004160, 140271929640448)
> >>> x[1] = 99
> >>> x
> [1, 99, 4]
> >>> y # 不受影响
> [1, 3, 4]
> ```

元组拆包

元组可以按如下顺序拆包赋值：

```
>>> a,b,c = 1,2,3
>>> a
1
>>> b
2
>>> c
3
```

Python 3 可以使用 * 运算符将元组进行块拆分。

```
>>> x,y,*z  = 1,2,3,4,5
>>> x
1
```

```
>>> y
2
>>> z
[3, 4, 5]
```

注意，变量 z 在赋值中包含了剩余的元素。你也可以改变分块的顺序。

```
>>> x,*y,z  = 1,2,3,4,5
>>> x
1
>>> y
[2, 3, 4]
>>> z
5
```

这种拆包有时被称为解构或展开。

字典

Python 字典是 Python 的核心，因为许多其他元素（例如函数、类）都是围绕它们构建的。有效的 Python 编程往往也意味着有效地使用字典。字典是实现映射数据结构的通用容器，有时称为哈希表或关联数组。字典需要一个键值（key/value）对，它将键映射到值。使用花括号 {} 和冒号 : 创建字典，格式如下：

```
>>> x = {'key': 'value'}
```

要从 x 字典中获取值，必须使用键来索引，如下所示：

```
>>> x['key']
'value'
```

让我们从一些基本语法开始。

```
>>> x={'play':'Shakespeare','actor':'Wayne','direct':'Kubrick',
...    'author':'Hemmingway','bio':'Watson'}

>>> len(x) # 键值对的数量
5
>>> x['pres']='F.D.R.' # 分配值给键 'pres'
>>> x
{'play': 'Shakespeare', 'actor': 'Wayne', 'direct': 'Kubrick',
↪ 'author': 'Hemmingway', 'bio': 'Watson', 'pres': 'F.D.R.'}
>>> x['bio']='Darwin' # 重新为键 'bio' 分配值
>>> x
{'play': 'Shakespeare', 'actor': 'Wayne', 'direct': 'Kubrick',
↪ 'author': 'Hemmingway', 'bio': 'Darwin', 'pres': 'F.D.R.'}
>>> del x['actor'] # 删除键值对
>>> x
{'play': 'Shakespeare', 'direct': 'Kubrick', 'author':
↪ 'Hemmingway', 'bio': 'Darwin', 'pres': 'F.D.R.'}
```

也可以使用内置函数 dict 创建字典。

```
>>> # 另一种创建字典的方法
>>> x=dict(key='value',
...        another_key=1.333,
...        more_keys=[1,3,4,'one'])
```

```
>>> x
{'key': 'value', 'another_key': 1.333, 'more_keys': [1, 3, 4,
↪   'one']}
>>> x={(1,3):'value'}    # 任何不可变类型都可以作为有效的键
>>> x
{(1, 3): 'value'}
>>> x[(1,3)]='immutables can be keys'
```

作为通用容器，字典可以包含其他字典、列表或其他 Python 类型。

> **编程技巧：字典的并集**
>
> 如果你想在一行代码中创建多个字典的并集，该怎么办？
> ```
> >>> d1 = {'a':1, 'b':2, 'c':3}
> >>> d2 = {'A':1, 'B':2, 'C':3}
> >>> dict(d1,**d2) # 用 dict 函数创建 d1 和 d2 的组合
> {'a': 1, 'b': 2, 'c': 3, 'A': 1, 'B': 2, 'C': 3}
> >>> {**d1,**d2} # 或直接这样创建
> {'a': 1, 'b': 2, 'c': 3, 'A': 1, 'B': 2, 'C': 3}
> ```

字典的不变性表现在只要提供一个有效的键就可以保证检索/存储到对应的值。请记住，列表是有序数据结构，列表元素在内存中是连续分布的。当列表元素被索引时，可以通过相对偏移量从前一个元素找到下一个元素。而字典则没有这个属性，因为它们将值放在它们能找到的内存中，无论是否连续。所以字典不依赖于相对偏移量进行索引，而是依赖于哈希函数。考虑以下内容：

```
>>> x = {0: 'zero', 1: 'one'}
>>> y = ['zero','one']
>>> x[1] # 字典
'one'
>>> y[1] # 列表
'one'
```

在上面的两种情况下，变量的索引方式看起来相同，但过程是不同的。当给定一个键时，字典计算哈希函数，并根据哈希函数在内存中存储值。什么是哈希函数？哈希函数接收一个输入，并以极高的概率返回唯一于该键的值。这意味着两个键不可能具有相同的哈希值，或者说，不能将不同的值存储在相同的内存位置中。

```
>>> hash('12345')
3973217705519425393
>>> hash('12346')
3824627720283660249
```

但是，这种唯一值的返回是有概率性的，因为内存是有限的，所以哈希函数也有很小的可能会产生相同的值，这被称为哈希冲突。对此问题，Python 有备用算法来处理这种情况。尽管如此，随着内存变得越来越稀缺，特别是在小

型平台上，如果你的代码使用许多大型字典，则寻找合适的内存块可能会变得很困难。

如之前讨论的那样，从列表中间插入/删除元素会导致额外的内存移动，因为列表维护其不变性，但这在字典中并不会发生。这意味着在字典中添加或删除元素不会有任何额外的内存开销。因此，对于不需要排序的代码，字典是理想的选择。请注意，自 Python 3.6+ 开始，字典是按照插入字典中的项的顺序排序的。在 Python 2.7 中，这被称为 collections.OrderedDict，但自 Python 3.6+ 开始已经成为默认设置。

在字典中，通常使用整数和字符串作为键，但任何不可变的类型都可以用作键，例如元组：

```
>>> x= {(1,3):10, (3,4,5,8):10}
```

如果使用可变类型作为键则会提示错误。

```
>>> a = [1,2]
>>> x[a]= 10
Traceback (most recent call last):
  File "<stdin>", line 1, in <module>
TypeError: unhashable type: 'list'
```

接下来看看为什么会发生这种情况。哈希函数保证在给定一个键时，它将始终能够检索到相应的值。假设在字典中可以使用可变的键，例如，假设 hash(a)->132334，并且将值 10 插入该内存中。如果在稍后的代码中更改了 a 的内容，例如 a[0]=3。那么由于哈希函数保证对不同的输入产生不同的输出，哈希函数的输出将与 132334 不同，因此字典无法检索相应的值，这将违反其不变性。所以，字典的键必须是不可变的。

集合

Python 提供了数学集合和相应的操作，通过 set() 数据结构实现，它们基本上是没有值的字典。

```
>>> set([1,2,11,1]) # 自动去重
{1, 2, 11}
>>> set([1,2,3]) & set([2,3,4]) # 交集
{2, 3}
>>> set([1,2,3]) and set([2,3,4])
{2, 3, 4}
>>> set([1,2,3]) ^ set([2,3,4]) # 两个集合中不同的元素
{1, 4}
>>> set([1,2,3]) | set([2,3,4]) # 两个集合并集的元素
{1, 2, 3, 4}
>>> set([ [1,2,3],[2,3,4] ]) # 不支持列表
```

```
(without more work)
Traceback (most recent call last):
  File "<stdin>", line 1, in <module>
TypeError: unhashable type: 'list'
```

注意，自 Python 3.6+ 开始，可以将字典的键用作集合对象，如下所示：

```
>>> d = dict(one=1,two=2)
>>> {'one','two'} & d.keys() # 交集
{'one', 'two'}
>>> {'one','three'} | d.keys() # 并集
{'one', 'two', 'three'}
```

哈希字典的值也可以作为集合对象。

```
>>> d = dict(one='ball',two='play')
>>> {'ball','play'} | d.items()
{'ball', 'play', ('one', 'ball'), ('two', 'play')}
```

创建集合后，可以按以下方式添加或删除单个元素：

```
>>> s = {'one',1,3,'10'}
>>> s.add('11')
>>> s
{1, 3, 'one', '11', '10'}
>>> s.discard(3)
>>> s
{1, 'one', '11', '10'}
```

请记住，集合是无序的，并且不能直接索引任何组成项。此外，subset( ) 方法用于真子集而不是部分子集。例如：

```
>>> a = {1,3,4,5}
>>> b = {1,3,4,5,6}
>>> a.issubset(b)
True
>>> a.add(1000)
>>> a.issubset(b)
False
```

issuperset 也是同理。在 Python 中，集合对于快速查找非常优化，如下所示：

```
>>> a = {1,3,4,5,6}
>>> 1 in a
True
>>> 11 in a
False
```

即使是对于大的集合，它的运行速度也非常快。

1.1.7　循环和条件

Python 中有两种主要的循环结构：for 循环和 while 循环。for 循环的语法很简单：

```
>>> for i in range(3):
...     print(i)
...
0
1
2
```

注意，for 循环语句的末尾有一个冒号字符，这是提示下一行应缩进。在 Python 中，块由空格缩进（推荐使用四个空格）表示，这使代码更易读。for 循环遍历迭代器中的项目，如上例中的 range(3)。Python 从循环结构中抽象出了可迭代的概念，因此有些 Python 对象本身就是可迭代的，只是需要 for 循环或 while 循环来执行。Python 有一个 else 子句，用于判断循环是否以 break㊀结束。

```
>>> for i in [1,2,3]:
...     if i>20:
...         break # 不会被执行
... else:
...     print('no break here!')
...
no break here!
```

只有当循环终止而没有中断时，才会执行 else 代码块。

while 循环也有类似的简单结构：

```
>>> i = 0
>>> while i < 3:
...     i += 1
...     print(i)
...
1
2
3
```

while 循环语句也有一个相应的可选 else 块。同样，冒号字符的存在提示下行需要缩进。while 循环会一直执行到布尔表达式（例如：i<3）的值为 False 为止。

逻辑和成员

Python 是一种真值语言，除以下情况外都为真：

- None
- False
- 任何数字类型的零，例如 0、0L、0.0、0j。
- 任何空序列，例如 "、( )、[ ]。
- 任何空映射，例如 { }。
- 如果定义了 _nonzero_( ) 或 _len_( ) 方法，而该方法返回整数零或布尔值 False。

㊀ continue 语句将跳转到 for 循环或 while 循环的顶部。

例如：

```
>>> bool(1)
True
>>> bool([]) # 空列表
False
>>> bool({}) # 空字典
False
>>> bool(0)
False
>>> bool([0,]) # 非空列表为真
True
```

Python 在处理数值区间的语法上非常简洁。

```
>>> 3.2 < 10 < 20
True
>>> True
True
```

可使用或（or）、非（not）和与（and）。

```
>>> 1 < 2 and 2 < 3 or 3 < 1
True
>>> 1 < 2 and not 2 > 3 or 1<3
True
```

使用分组括号来增加代码的可读性是非常重要的。在 Python 中，你还可以对可迭代对象进行逻辑操作，如下所示：

```
>>> (1,2,3) < (4,5,6) # 至少一个为真
True
>>> (1,2,3) < (0,1,2) # 全部为假
False
```

在 Python 中处理字符串时，应该避免使用相对比较操作符（例如 'a' < 'b'），因为这样的代码可读性较差。相反，应该使用字符串匹配操作符（例如，==）来进行字符串的比较。成员身份测试可以使用关键字 in：

```
>>> 'on' in [22,['one','too','throw']]
False
>>> 'one' in [22,['one','too','throw']] # 无递归
False
>>> 'one' in [22,'one',['too','throw']]
True
>>> ['too','throw'] not in [22,'one',['too','throw']]
False
```

如果在数百万个元素中测试成员身份，则使用 set() 比使用列表快得多。例如：

```
>>> 'one' in {'one','two','three'}
True
```

用"is"关键字检查两个对象是否相同。

```
>>> x = 'this string'
>>> y = 'this string'
>>> x is y
```

```
False
>>> x==y
True
```

注意，用"is"关键字检查的是每个对象的标识符（id），例如：

```
>>> x=y='this string'
>>> id(x),id(y)
(140271930045360, 140271930045360)
>>> x is y
True
```

"is"还可以有以下用法：x is True，x is None。

注意，None 是 Python 的单例对象。

条件

使用 if 来构建条件语句。

```
>>> if 1 < 2:
...     print('one less than two')
...
one less than two
```

在 Python 中，支持"else"和"elif"与"if"一起使用构建条件语句，但不支持"switch"。

```
>>> a = 10
>>> if a < 10:
...     print('a less than 10')
... elif a < 20:
...     print('a less than 20')
... else:
...     print('a otherwise')
...
a less than 20
```

还有一种不太常见的单行语法也可以用于条件判断。

```
>>> x = 1 if (1>2) else 3 # 单行条件
>>> x
3
```

列表解析

在 Python 中，采用循环语句收集 items 是一种常见的操作，惯用语法如下：

```
>>> out=[] # 初始化
>>> for i in range(10):
...     out.append(i**2)
...
>>> out
[0, 1, 4, 9, 16, 25, 36, 49, 64, 81]
```

上面的代码可以缩写为列表解析。

```
>>> [i**2 for i in range(10)] # 整数的二次方
[0, 1, 4, 9, 16, 25, 36, 49, 64, 81]
```

列表解析中也可以包含条件。

```
>>> [i**2 for i in range(10) if i % 2] # 包含条件
[1, 9, 25, 49, 81]
```

这相当于:

```
>>> out = []
>>> for i in range(10):
...     if i % 2:
...         out.append(i**2)
...
>>> out
[1, 9, 25, 49, 81]
```

解析也适用于字典和集合。

```
>>> {i:i**2 for i in range(5)} # 字典
{0: 0, 1: 1, 2: 4, 3: 9, 4: 16}
>>> {i**2 for i in range(5)}    # 集合
{0, 1, 4, 9, 16}
```

1.1.8 函数

定义函数有两种常见的方法,以下使用关键字 "def" 定义函数:

```
>>> def foo():
...     return 'I said foo'
...
>>> foo()
'I said foo'
```

定义函数时需要使用 return 语句。因为如果没有 return 语句,函数只会返回 None。而在调用函数时,需要使用到括号,因为在 Python 中函数属于一级对象。函数作为一级对象可以像任何其他 Python 对象一样进行操作,它们可以放在容器中并传递,而无需任何特殊处理。

```
>>> foo # 无括号,则此处仅是一个 Python 对象
<function foo at 0x7f939a6ecc10>
```

在 Python 的早期,这是一个关键特性,否则只能通过指针传递函数,而这需要特殊处理。实际上,一级对象可以像其他 Python 对象一样操作,它们可以被放入容器中,无需特殊处理即可传递。当然,我们希望向函数提供参数。函数的参数有两种类型:位置参数和关键字参数。

```
>>> def foo(x): # 位置参数
...     return x*2
...
>>> foo(10)
20
```

位置参数可以指定其名称。

```
>>> def foo(x,y):
...     print('x=',x,'y=',y)
...
```

变量 x 和 y 可以通过其位置指定，如下所示：

```
>>> foo(1,2)
x= 1 y= 2
```

位置参数也可以使用名称而不是位置来指定，例如：

```
>>> foo(y=2,x=1)
x= 1 y= 2
```

关键字参数允许指定默认值。

```
>>> def foo(x=20): # 关键字参数
...     return 2*x
...
>>> foo(1)
2
>>> foo()
40
>>> foo(x=30)
60
```

函数可以有多个设有默认值的关键字参数。

```
>>> def foo(x=20,y=30):
...     return x+y
...
>>> foo(20,)
50
>>> foo(1,1)
2
>>> foo(y=12)
32
>>> help(foo)
Help on function foo:

foo(x=20, y=30)
```

> **编程技巧：函数的文档**
>
> 在编写函数时，应该尽可能使用有意义的变量名和默认值，并为其添加文档字符串。这样可以使你的代码易于导航、理解和使用。

在 Python 中可以使用文档字符串（docstrings）来为函数提供文档注释，以便在调用 help 来了解函数时能获得更多关于函数的信息，例如：

```
>>> def compute_this(position=20,velocity=30):
...     '''position in m
...     velocity in m/s
...     '''
...     return x+y
...
>>> help(compute_this)
Help on function compute_this:

compute_this(position=20, velocity=30)
    position in m
    velocity in m/s
```

因此，通过使用有意义的参数和函数名，并在 docstrings 中包含基本文档，可以极大地提高 Python 函数的可用性。此外，建议将函数名设置为动词形式（例如，get_this, compute_field）。

除了使用"def"定义函数，还可以使用 lambda 创建单行函数。这样申明的函数有时称为匿名函数。

```
>>> f = lambda x:  x**2 # 匿名函数
>>> f(10)
100
```

作为一级对象，函数可以像任何其他 Python 对象一样放入列表。

```
>>> [lambda x: x, lambda x:x**2] # 函数列表
[<function <lambda> at 0x7f939a6ba700>, <function <lambda> at
↪   0x7f939a6ba790>]
>>> for i in  [lambda x: x, lambda x:x**2]:
...    print(i(10))
...
10
100
```

目前为止，我们还没有充分利用元组（tuple），但这种数据结构与函数一起使用时会变得非常强大。这是因为元组允许将函数参数与函数本身分离。这意味着可以在不同地方传递元组，构建函数参数，并在之后使用一个或多个函数执行它们。示例如下：

```
>>> def foo(x,y,z):
...    return x+y+z
...
>>> foo(1,2,3)
6
>>> args = (1,2,3)
>>> foo(*args) # 单星号,使用元组传递参数
6
```

元组前面的星号符号会将元组解包到函数签名中。我们已经在元组的解包赋值中见过这种用法。

```
>>> x,y,z = args
>>> x
1
>>> y
2
>>> y
2
```

双星号符号则用于对关键字参数进行相应的解包，这与在函数签名中的解包操作相同。

```
>>> def foo(x=1,y=1,z=1):
...    return x+y+z
...
>>> kwds = dict(x=1,y=2,z=3)
```

```
>>> kwds
{'x': 1, 'y': 2, 'z': 3}
>>> foo(**kwds) # 双星号,使用字典传递参数
6
```

使用元组和字典同时传递参数，示例如下：

```
>>> def foo(x,y,w=10,z=1):
...     return (x,y,w,z)
...
>>> args = (1,2)
>>> kwds = dict(w=100,z=11)
>>> foo(*args,**kwds)
(1, 2, 100, 11)
```

函数变量作用域

函数或子函数中的变量是各自作用域的局部变量。如果要在函数内对全局变量进行更改，则需要进行特殊的操作。

```
>>> x=10 # 函数外
>>> def foo():
...     return x
...
>>> foo()
10
>>> print('x = %d is not changed'%x)
x = 10 is not changed

>>> def foo():
...     x=1 # 函数内部定义的变量
...     return x
...
>>> foo()
1
>>> print('x = %d is not changed'%x)
x = 10 is not changed

>>> def foo():
...     global x # 定义为全局变量
...     x=20     # 函数内赋值
...     return x
...
>>> foo()
20
>>> print('x = %d IS changed!'%x)
x = 20 is changed!
```

函数关键字过滤

在函数定义中使用**kwds参数可以使函数忽略未使用的关键字参数。

```
>>> def foo(x=1,y=2,z=3,**kwds):
...     print('in foo, kwds = %s'%(kwds))
...     return x+y+z
...
>>> def goo(x=10,**kwds):
...     print('in goo, kwds = %s'%(kwds))
...     return foo(x=2*x,**kwds)
...
```

```
>>> def moo(y=1,z=1,**kwds):
...     print('in moo, kwds = %s'%(kwds))
...     return goo(x=z+y,z=z+1,q=10,**kwds)
...
```

可以使用未指定的关键字参数调用其中的任何一个函数，如下所示：

```
>>> moo(y=91,z=11,zeta_variable = 10)
in moo, kwds = {'zeta_variable': 10}
in goo, kwds = {'z': 12, 'q': 10, 'zeta_variable': 10}
in foo, kwds = {'q': 10, 'zeta_variable': 10}
218
```

在上面的示例中，zeta_variable 变量作为未使用的关键字参数在函数间被传递。使用 **kwds 进行关键字过滤，可以使未使用的关键字参数在函数间得到有效传递。这是一个非常有用的 Python 特性，可用于进行代码包装。

由于这是一个非常出色且实用的 Python 特性，下面再提供一个示例，我们可以在其中追踪每个函数签名是如何被满足的，以及剩余的关键字参数是如何被传递的。

```
>>> def foo(x=1,y=2,**kwds):
...     print('foo: x = %d, y = %d, kwds=%r'%(x,y,kwds))
...     print('\t',)
...     goo(x=x,**kwds)
...
>>> def goo(x=10,**kwds):
...     print('goo: x = %d, kwds=%r'%(x,kwds))
...     print('\t\t',)
...     moo(x=x,**kwds)
...
>>> def moo(z=20,**kwds):
...     print('moo: z=%d, kwds=%r'%(z,kwds))
...
```

然后，

```
>>> foo(x=1,y=2,z=3,k=20)
foo: x = 1, y = 2, kwds={'z': 3, 'k': 20}

goo: x = 1, kwds={'z': 3, 'k': 20}

moo: z=3, kwds={'x': 1, 'k': 20}
```

需注意每个函数的函数签名是如何被满足的，以及剩余的关键字参数是如何被传递的。Python 3 中可以通过 * 符号强制用户提供关键字参数，示例如下：

```
>>> def foo(*,x,y,z):
...     return x*y*y
...
```

然后，

```
>>> foo(1,2,3)          # 没有必需的关键字参数吗
Traceback (most recent call last):
  File "<stdin>", line 1, in <module>
TypeError: foo() takes 0 positional arguments but 3 were given
>>> foo(x=1,y=2,z=3) # 必须提供关键字参数
4
```

函数中虽然可以使用 * args 和 ** kwargs 提供的通用接口来处理函数参数，但是这些参数并不适用于集成代码开发工具，因为变量自省（introspection）对这些函数签名不起作用。例如：

```
>>> def foo(*args,**kwargs):
...     return args, kwargs
...
>>> foo(1,2,2,x=12,y=2,q='a')
((1, 2, 2), {'x': 12, 'y': 2, 'q': 'a'})
```

这样做会将参数的处理交给函数本身，这会导致函数签名不清晰。因此，应尽可能避免这种情况。

函数式编程惯用语法

虽然 Python 不是像 Haskell 一样的真正的函数式编程语言，但它也有一些有用的函数式惯用语法。这些惯用语法在并行计算框架（如 PySpark）中变得非常重要。例如，使用 map 函数将给定的（lambda）函数应用于 range(10) 中的每个可迭代对象：

```
>>> map(lambda x:  x**2 , range(10))
<map object at 0x7f939a6a6fa0>
```

想要获得输出结果，则需要将其转换为列表，示例如下：

```
>>> list(map(lambda x:  x**2, range(10)))
[0, 1, 4, 9, 16, 25, 36, 49, 64, 81]
```

这与相应的列表输出相同。

```
>>> list(map(lambda x:  x**2, range(10)))
[0, 1, 4, 9, 16, 25, 36, 49, 64, 81]
>>> [i**2 for i in range(10)]
[0, 1, 4, 9, 16, 25, 36, 49, 64, 81]
```

还有 reduce 函数：

```
>>> from functools import reduce
>>> reduce(lambda x,y:x+2*y,[0,1,2,3],0)
12
>>> [i for i in range(10) if i %2 ]
[1, 3, 5, 7, 9]
```

用 functools.reduce 解决递归问题的效率非常高，例如最小公倍数算法的实现：

```
>>> from functools import reduce
>>> def gcd(a, b):
...     'Return greatest common divisor using Euclids Algorithm.'
...     while b:
...         a, b = b, a % b
...     return a
...
>>> def lcm(a, b):
...     'Return lowest common multiple.'
```

```
...     return a * b // gcd(a, b)
...
>>> def lcmm(*args):
...     'Return lcm of args.'
...     return reduce(lcm, args)
...
```

> **编程技巧：注意函数中的默认容器**
> ```
> >>> def foo(x=[]):  # 使用空列表作为默认值
> ...     x.append(10)
> ...     return x
> ...
> >>> foo()  # 也许你期望得到这个...
> [10]
> >>> foo()  # 但你是否期望得到这个...
> [10, 10]
> >>> foo()  # ...还是这个？这里发生了什么？
> [10, 10, 10]
> ```

未指定的参数通常在函数签名中使用 None 来处理。

```
>>> def foo(x=None):
...     if x is None:
...         x = 10
...
```

代码逻辑解决了缺失的项目。

深入理解 Function

让我们通过一个示例来了解 Python 如何构造 Function 对象。

```
>>> def foo(x):
...     return x
...
```

foo 函数的内部信息可以通过 __code__ 查询。例如，foo.__code__.co_argcount 给出 foo 函数的参数数量：

```
>>> foo.__code__.co_argcount
1
```

co_varnames 属性返回一个元组，包含了函数中定义的所有变量名。

```
>>> foo.__code__.co_varnames
('x',)
```

函数的局部变量也可以通过 __code__ 查询到。

```
>>> def foo(x):
...     y = 2*x
...     return y
...
>>> foo.__code__.co_varnames
('x', 'y')
```

函数还可以使用 *args 来表示在定义函数时未指定的任意输入。

```
>>> def foo(x,*args):
...     return x+sum(args)
...
>>> foo.__code__.co_argcount # 和之前一样吗?
1
```

在上面这种情况中,因为*args是用函数对象的co_flags属性处理的,所以foo.__code__.co_argcount返回的函数参数数量依旧为1。

```
>>> print('{0:b}'.format(foo.__code__.co_flags))
1000111
```

注意,foo.__code__.co_flags返回的位掩码中,第3个比特位为1,表示函数签名包含*args。在十六进制中,0x01掩码对应于co_optimized,0x02对应于co_newlocals,0x04对应于co_varargs,0x08对应于co_varkeywords,0x10对应于co_nested,0x20对应于co_generator。

另外,可以使用dis模块对函数对象进行解包。

```
>>> def foo(x):
...     y= 2*x
...     return y
...
>>> import dis
>>> dis.show_code(foo)
Name:              foo
Filename:          <stdin>
Argument count:    1
Positional-only arguments: 0
Kw-only arguments: 0
Number of locals:  2
Stack size:        2
Flags:             OPTIMIZED, NEWLOCALS, NOFREE
Constants:
   0: None
   1: 2
Variable names:
   0: x
   1: y
```

注意,上面的代码中,常量(constants)没有被编译进字节码(byte-code)中,而是存储在函数对象中。foo.__code__.co_consts给出foo函数对象的常量,例如:

```
>>> def foo(x):
...     a,b = 1,2
...     return x*a*b
...
>>> print(foo.__code__.co_varnames)
('x', 'a', 'b')
>>> print(foo.__code__.co_consts)
(None, (1, 2))
```

如果需要更全面地了解函数对象,还可以使用dis.dis更详细地检查co_code信息。

```
>>> print(foo.__code__.co_code)  # 原始字节码
b'd\x01\\\x02}\x01}\x02|\x00|\x01\x14\x00|\x02\x14\x00S\x00'
>>> dis.dis(foo)
  2           0 LOAD_CONST               1 ((1, 2))
              2 UNPACK_SEQUENCE          2
              4 STORE_FAST               1 (a)
              6 STORE_FAST               2 (b)

  3           8 LOAD_FAST                0 (x)
             10 LOAD_FAST                1 (a)
             12 BINARY_MULTIPLY
             14 LOAD_FAST                2 (b)
             16 BINARY_MULTIPLY
             18 RETURN_VALUE
```

其中，LOAD_CONST 为函数中存储的常数，LOAD_FAST 为函数中的局部变量。函数对象将默认值存储在元组 __defaults__ 中。例如：

```
>>> def foo(x=10):
...     return x*10
...
>>> print(foo.__defaults__)
(10,)
```

函数的作用域遵循 LEGB 原则，即局部（Local）、嵌套（Enclosing）、全局（Global）、内置（Built-ins）的顺序。在函数体中，当 Python 遇到一个变量时，首先检查它是否为局部变量（co_varnames），如果不是再检查其他作用域。这里有一个有趣的例子。

```
>>> def foo():
...     print('x=',x)
...     x = 10
...
>>> foo()
Traceback (most recent call last):
  File "<stdin>", line 1, in <module>
  File "<stdin>", line 2, in foo
UnboundLocalError: local variable 'x' referenced before assignment
```

为什么会这样？让我们看看函数内部。

```
>>> foo.__code__.co_varnames
('x',)
```

当 Python 尝试解析 x 时，它会检查其是否为局部变量，这是因为它出现在 co_varnames 中，但没有对其赋值，因此会生成 UnboundLocalError。

函数作用域的嵌套作用域更为复杂，例如：

```
>>> def outer():
...     a,b = 0,0
...     def inner():
...         a += 1
...         b += 1
...         print(f'{a},{b}')
...     return inner
...
```

```
>>> f = outer()
>>> f()
Traceback (most recent call last):
  File "<stdin>", line 1, in <module>
  File "<stdin>", line 4, in inner
UnboundLocalError: local variable 'a' referenced before assignment
```

可以看到上面出现了错误，让我们再来检查一下函数对象。

```
>>> f.__code__.co_varnames
('a', 'b')
```

这意味着，内部函数认为这些变量是局部变量。但实际上这些变量存在于嵌套作用域中。可以用关键字 nonlocal 来解决这个问题，例如：

```
>>> def outer():
...     a,b = 0,0
...     def inner():
...         nonlocal a,b # 使用 nonlocal 来申明非局部变量
...         a+=1
...         b+=1
...         print(f'{a},{b}')
...     return inner
...
>>> f = outer()
>>> f()
1,1
```

如果回头检查 co_varnames 属性，你会发现它是空的。内部函数中的 co_freevars 包含了非局部变量的信息。

```
>>> f.__code__.co_freevars
('a', 'b')
```

这样内部函数就知道这些变量在封闭作用域中是什么。外部函数也通过 co_cellvars 属性知道嵌套函数中使用了哪些变量。

```
>>> outer.__code__.co_cellvars
('a', 'b')
```

因此，这种封闭关系是双向的。

函数栈帧

嵌套的 Python 函数放在一个堆栈上，举例如下：

```
>>> def foo():
...     return 1
...
>>> def goo():
...     return 1+foo()
...
```

当调用 goo 时，它首先调用 foo，因为 foo 位于堆栈中的 goo 之上。堆栈框架是一种数据结构，用于维护程序作用域和有关执行状态的信息。因此，在本例中，每个函数有两个帧，其中 foo 位于堆栈的最顶层。

可以使用以下方法通过堆栈框架检查内部执行状态：

```
>>> import sys
>>> depth = 0 # 栈顶
>>> frame = sys._getframe(depth)
>>> frame
<frame at 0x7f939a6bcba0, file '<stdin>', line 1, code <module>>
```

其中，depth 是相对于堆栈顶部的调用数，0 对应于当前帧。帧包含当前范围中的局部变量（frame.f_locals）和当前模块中的全局变量（frame.f_globals）。请注意，也可以使用 locals ( ) 和 globals ( ) 内置函数访问局部和全局变量。帧对象也具有 f_lineno（当前行号）、f_trace（跟踪函数）和 f_back（引用前一帧）属性。在一个未处理的异常中，Python 使用 f_back 向后导航以显示堆栈帧。删除帧对象很重要，否则会有创建循环引用的危险。

堆栈帧还包含 frame.f_code，它是可执行的字节码。这与函数对象不同，因为它不包含对全局执行环境的引用。例如，可以使用内置的 compile 函数创建代码对象。

```
>>> c = compile('1 + a*b','tmp.py','eval')
>>> print(c)
<code object <module> at 0x7f939a6a5df0, file "tmp.py", line 1>
>>> eval(c,dict(a=1,b=2))
3
```

注意，eval 会评估代码对象的表达式。代码对象包含 co_filename（创建的文件名）、co_name（函数/模块名）、co_varnames（变量名）和 co_code（编译后的字节码）属性。

> **编程技巧：函数中的断言**
>
> 编写干净且可重用的函数是 Python 编程的基础。大幅提高函数可靠性和可重用性的最简单方法是在代码中加入断言（assert）语句。如果这些语句为 False，则会引发断言错误（AssertionError），并提供了一种很好的方法来确保函数按预期运行。考虑以下示例：
>
> ```
> >>> def foo(x):
> ...     return 2*x
> ...
> >>> foo(10)
> 20
> >>> foo('x')
> 'xx'
> ```
>
> 若定义 foo 函数的目的是处理数值型输入，则 foo('x') 的正常执行就属于"故障穿透"现象。"故障穿透"是由 Python 的动态类型特性所带来的隐蔽性缺陷。要想解决这个问题，一种快速的方法是使用断言，如下所示：

```
>>> def foo(x):
...     assert isinstance(x,int)
...     return 2*x
...
>>> foo('x')
Traceback (most recent call last):
  File "<stdin>", line 1, in <module>
  File "<stdin>", line 2, in foo
AssertionError
```

现在，该函数仅限于使用整数输入，并且如果输入不符合要求则会引发 AssertionError。除了检查类型外，assert 语句还可以通过检查中间计算来确保函数的业务逻辑。此外，Python 中有一个命令行开关可以关闭 assert 语句（如果需要），但是 assert 语句不应过于复杂，而且应该为意外用法提供备选方案。

清晰且富有表现力的函数是精通 Python 编程的重要标志。函数的用户会受益于清晰的变量和函数名以及详细文档。有一个判断代码好坏的经验是：说明文档 docstring 应该长于函数的主体。如果不是，那就应该将函数切分成更小的模块。有关 Python 代码的示例，可查看 networkx[①]项目。

惰性求值和函数签名

考虑以下函数：
```
>>> def foo(x):
...     return a*x
...
```

注意，定义这个函数时，即使变量 a 未定义，解释器也不会报错，只有当你尝试运行函数时才会报错。这是因为 Python 函数是惰性求值的，这意味着函数体只有在被调用时才会被处理。因此，函数尝试在命名空间中查找缺失的变量 a 之前并不知道它存在与否。尽管函数体是惰性求值的，但函数签名是立即求值的。

```
>>> def report():
...     print('I was called!')
...     return 10
...
>>> def foo(x = report()):
...     return x
...
I was called!
```

注意，上面的代码中 foo 没有被调用，但 report() 被调用了，因为它出现在 foo 的函数签名中。

---

① 见 https://networkx.org/。

1.1.9 文件输入／输出

在 Python 中，文件的读写非常简单直接，方法如下：

```
>>> f=open('myfile.txt','w') # 写模式
>>> f.write('this is line 1')
14
>>> f.close()
>>> f=open('myfile.txt','a') # append 追加模式
>>> f.write('this is line 2')
14
>>> f.close()
>>> f=open('myfile.txt','a') # append 追加模式
>>> f.write('\nthis is line 3\n')  # 写入新的一行
16
>>> f.close()
>>> f=open('myfile.txt','a') # append 追加模式
>>> f.writelines([ 'this is line 4\n', 'this is line 5\n'])  #
↪   写入新的一行
>>> f=open('myfile.txt','r') # 读模式
>>> print(f.readlines())
['this is line 1this is line 2\n', 'this is line 3\n', 'this is
↪   line 4\n', 'this is line 5\n']
>>> ['this is line 1this is line 2\n', 'this is line 3\n', 'this
↪   is line 4\n', 'this is line 5\n']
['this is line 1this is line 2\n', 'this is line 3\n', 'this is
↪   line 4\n', 'this is line 5\n']
```

可以使用类似 seek() 的方法在文件中移动指针。无需手动关闭文件，with 语句会自动完成这项操作。示例如下：

```
>>> with open('myfile.txt','r') as f:
...     print(f.readlines())
...
['this is line 1this is line 2\n', 'this is line 3\n', 'this is
↪   line 4\n', 'this is line 5\n']
```

with 语句的主要优点是文件将在块结束时自动关闭，无需专门使用 f.close 手动关闭。简而言之，当输入 with 块时，Python 运行 f.__enter__ 方法打开文件，然后在块的末尾运行 f.__exit__ 方法。with 语句可用于其他符合此协议的对象和上下文管理器[○]。

注意，当写入非文本文件，应使用 'rb' 和 'wb' 来代替 'r' 和 'w'。

> **编程技巧：其他 I/O 模块**
>
> Python 有许多工具用于处理不同级别的文件 I/O。struct 模块适用于纯二进制读写。mmap 模块对于绕过文件系统使用虚拟内存进行快速文件访问非常有用。StringI/O 模块允许字符串读写。

○ 见 Contextlib 内置模块。

序列化：保存复杂对象

序列化意味着打包 Python 对象，以便在不同的 Python 进程或不同的计算机之间传送。Python 的多平台特性意味着无法确保 Python 对象的低级属性（例如，平台类型或 Python 版本之间）保持一致。

在绝大多数情况下，可以使用 pickle 模块来序列化，示例如下：

```
>>> import pickle
>>> mylist = ["This", "is", 4, 13327]
>>> f=open('myfile.dat','wb') # 二进制模式
>>> pickle.dump(mylist, f)
>>> f.close()
>>> f=open('myfile.dat','rb') # 写入模式
>>> print(pickle.load(f))
['This', 'is', 4, 13327]
```

pickle 模块实现了对一个 Python 对象结构的二进制序列化和反序列化，在 pickle 中使用 dump（ ） 和 load（ ） 读写文件时，要使用 'rb' 或 'wb' 模式。

> **编程技巧：通过套接字进行序列化**
>
> 可以通过创建中间文件、使用套接字或其他协议来序列化字符串并进行传输。

序列化函数

函数的内部状态以及它如何连接到创建它的 Python 进程，使得对函数进行序列化变得棘手。有一种方法是使用 marshal 模块将函数对象以二进制格式转储，将其写入文件，然后在另一端使用 types.FunctionType 进行重建。这种技术的缺点是，虽然它们都是 CPython 实现，但在不同的 Python 版本之间可能不兼容。

```
>>> import marshal
>>> def foo(x):
...     return x*x
...
>>> code_string = marshal.dumps(foo.__code__)
```

在远端进程中（在传输 code_string 后）：

```
>>> import marshal, types
>>> code = marshal.loads(code_string)
>>> func = types.FunctionType(code, globals(), "some_func_name")
>>> func(10)   # gives 100
100
```

> **编程技巧：Dill Pickles**
>
> dill 模块可以对函数进行序列化并处理所有这些复杂情况。然而，一旦导入 dill，所有的序列化都将被 dill 接管。如果想要在导入 dill 后使用 dill 进行更

精细的序列化控制而不被接管，可以在导入 dill 后执行 dill.extend(False)。

```
import dill
def foo(x):
    return x*x

dill.dumps(foo)
```

1.1.10　处理错误

Python 最有魅力的地方是它允许先行动再请求原谅，例如：

```
try:
    # 尝试一些操作
except:
    # 解决一些问题
```

上面的 except 块将能捕获并处理在 try 块中抛出的任何类型的异常。Python 提供了一长串内置异常，你可以捕获它们，也可以根据需要创建自己的异常。除了捕获异常，还可以使用 raise 语句引发自己的异常。此外，Python 还提供了一个 assert 语句，如果某些条件在断言时不为真，该语句会抛出异常。

以下是 Python 异常处理功能的一些示例：

```
>>> def some_function():
...     try:
...         # 除以 0 会引发一个异常
...         10 / 0
...     except ZeroDivisionError:
...         print("Oops, invalid.")
...     else:
...         # 异常没有发生
...         pass
...     finally:
...         # 此处在代码块执行完毕并且所有异常被处理后执行
...         print("We're done with that.")
...
>>> some_function()
Oops, invalid.
We're done with that.
>>> out = list(range(3))
```

异常可以非常具体化，例如：

```
>>> try:
...     10 / 0
... except ZeroDivisionError:
...     print('I caught an attempt to divide by zero')
...
I caught an attempt to divide by zero
>>> try:
...     out[999]  # 引发 IndexError 异常
```

```
... except ZeroDivisionError:
...     print('I caught an attempt to divide by zero')
...
Traceback (most recent call last):
  File "<stdin>", line 2, in <module>
IndexError: list index out of range
```

捕捉到的异常会改变代码流。

```
>>> try:
...     1/0
...     out[999]
... except ZeroDivisionError:
...     print('I caught an attempt to divide by zero but I did not
↪  try out[999]')
...
I caught an attempt to divide by zero but I did not try out[999]
```

try 块中异常的顺序很重要。

```
>>> try:
...     1/0         # 引发 ZeroDivisionError 异常
...     out[999]    # 永远不会执行到这里
... except IndexError:
...     print('I caught an attempt to index something out of
↪  range')
...
Traceback (most recent call last):
  File "<stdin>", line 2, in <module>
ZeroDivisionError: division by zero
```

块可以嵌套，但如果这些块的深度超过两层，则对代码整体来说就不太好了。

```
>>> try:    # 嵌套的异常
...     try: # 内部
...         1/0
...     except IndexError:
...         print('caught index error inside')
... except ZeroDivisionError as e:
...     print('I caught an attempt to divide by zero inside
↪  somewhere')
...
I caught an attempt to divide by zero inside somewhere
```

嵌套中使用 finally 子句。

```
>>> try:    # 嵌套的异常
...     try:
...         1/0
...     except IndexError:
...         print('caught index error inside')
...     finally:
...         print("I am working in inner scope")
... except ZeroDivisionError as e:
...     print('I caught an attempt to divide by zero inside
↪  somewhere')
...
I am working in inner scope
I caught an attempt to divide by zero inside somewhere
```

异常可以按元组的形式进行分组。

```
>>> try:
...     1/0
... except (IndexError,ZeroDivisionError) as e:
...     if isinstance(e,ZeroDivisionError):
...         print('I caught an attempt to divide by zero inside
 somewhere')
...     elif isinstance(e,IndexError):
...         print('I caught an attempt to index something out of
 range')
...
I caught an attempt to divide by zero inside somewhere
```

尽管可以通过一个未指定类型的 except 行来捕获任何异常，但这样做无法告诉你抛出了什么异常，所以可以使用 Exception 来揭示异常信息。

```
>>> try: # 更详细的异常捕获
...     1/0
... except Exception as e:
...     print(type(e))
...
<class 'ZeroDivisionError'>
```

> **编程技巧：使用 Exceptions 控制函数递归**
>
> 嵌套函数调用会导致更多的堆栈帧，最终达到递归极限。例如，下面的递归函数在 n 足够大的情况下最终会失败。
>
> ```
> >>> def factorial(n):
> ...     if (n == 0): return 1
> ...     return n * factorial(n-1)
> ...
> ```
>
> Python 的 exceptions 可用来阻止额外的帧堆积在堆栈。下面的递归是 Exception 的一个子类，只做保存参数的事情。递归函数会引发递归异常，该异常在调用时堆栈停止增长。这两个定义的目的是创建 tail_recursive 递归装饰器。该装饰器返回一个嵌入无限 while 循环的函数，该循环只有在此装饰器终止而不触发额外的递归步骤时才能退出。否则，被装饰函数的输入参数被传递，而输入参数实际上是递归计算的中间值。该技术绕过了 Python 内置的递归限制，可用作任何使用输入参数存储中间值的递归函数的装饰器。[1]
>
> ```
> >>> class Recurse(Exception):
> ...     def __init__(self, *args, **kwargs):
> ...         self.args, self.kwargs = args, kwargs
> ...
> >>> def recurse(*args, **kwargs):
> ...     raise Recurse(*args, **kwargs)
> ...
> ```

---

[1] 可参见 https://chrispenner.ca/posts/python-tail-recursion 中对这种技术的深入讨论。

```
>>> def tail_recursive(f):
...     def decorated(*args, **kwargs):
...         while True:
...             try:
...                 return f(*args, **kwargs)
...             except Recurse as r:
...                 args, kwargs = r.args, r.kwargs
...     return decorated
...
>>> @tail_recursive
... def factorial(n, accumulator=1):
...     if n == 0: return accumulator
...     recurse(n-1, accumulator=accumulator*n)
...
```

1.1.11 掌握 Python 的强大功能

下面这些 Python 内置代码的熟练使用将有助于提高 Python 编程水平。

zip 函数

Python 有一个内置的 zip 函数，可以将可迭代的对象成对组合。

```
>>> zip(range(3),'abc')
<zip object at 0x7f939a91d200>
>>> list(zip(range(3),'abc'))
[(0, 'a'), (1, 'b'), (2, 'c')]
>>> list(zip(range(3),'abc',range(1,4)))
[(0, 'a', 1), (1, 'b', 2), (2, 'c', 3)]
```

还可以使用 * 来反转这一操作。

```
>>> x = zip(range(3),'abc')
>>> i,j = list(zip(*x))
>>> i
(0, 1, 2)
>>> j
('a', 'b', 'c')
```

当与 dict 结合使用时，zip 提供了一种构建 Python 字典的强大方法。

```
>>> k = range(5)
>>> v = range(5,10)
>>> dict(zip(k,v))
{0: 5, 1: 6, 2: 7, 3: 8, 4: 9}
```

max 函数

max 函数用来取序列的最大值。

```
>>> max([1,3,4])
4
```

如果序列中的项是元组，则元组中的第一项用于排序。

```
>>> max([(1,2),(3,4)])
(3, 4)
```

max 函数还可以通过关键字参数控制如何计算序列中的项。例如，可以根据元组中的第二个元素进行排序：

```
>>> max([(1,4),(3,2)], key=lambda i:i[1])
(1, 4)
```

关键字参数也适用于 min 和 sorted 函数。

with 语句

with 语句为后续代码设置上下文。

```
>>> class ControlledExecution:
...     def __enter__(self):
...         # 设置
...         print('I am setting things up for you!')
...         return 'something to use in the with-block'
...     def __exit__(self, type, value, traceback):
...         # 清理资源
...         print('I am tearing things down for you!')
...
>>> with ControlledExecution() as thing:
...     # 代码
...     pass
...
I am setting things up for you!
I am tearing things down for you!
```

用 with 语句打开和关闭文件。

```
f = open("sample1.txt") # 文件句柄
f.__enter__()
f.read(1)
f.__exit__(None, None, None)
f.read(1)
```

这种方法可以防止忘记关闭文件。

```
with open("x.txt") as f:
    data = f.read()
    # 对数据进行处理
```

contextlib 快速上下文管理模块

contextlib 模块可以方便快速地创建上下文管理器。

```
>>> import contextlib
>>> @contextlib.contextmanager
... def my_context():
...     print('setting up ')
...     try:
...         yield {} # 如果需要,可以为'as'部分生成对象
...     except:
...         print('catch some errors here')
...     finally:
...         print('tearing down')
...
>>> with my_context():
...     print('I am in the context')
...
```

```
setting up
I am in the context
tearing down
>>> with my_context():
...     raise RuntimeError ('I am an error')
...
setting up
catch some errors here
tearing down
```

Python 内存管理

在之前的学习中,我们曾使用 id 函数获取 Python 对象的唯一标识符。在 CPython 实现中,这实际上是对象的内存位置。Python 中的任何对象都有一个引用计数器,用于跟踪指向它的标签。我们在前面讨论具有不同标签(即变量名)的列表时看到了这一点。当不再有指向给定对象的标签时,该对象的内存将被释放。但是在有循环引用时,这种方式就会遇到问题:

```
>>> x = [1,2]
>>> x.append(x) # 循环引用
>>> x
[1, 2, [...]]
```

在这种情况下,引用计数器永远不会倒计时到零并被释放。为了防止这种情况,Python 采用了垃圾收集器,它能定期停止执行的主线程,以查找和删除任何此类引用。

下面的代码使用功能强大的 ctypes 模块,它可以直接访问 Cpython 中 object.h 文件的 C 结构体中的引用计数器字段。

```
>>> def get_refs(obj_id):
...     from ctypes import Structure, c_long
...     class PyObject(Structure):
...         _fields_ = [("reference_cnt", c_long)]
...     obj = PyObject.from_address(obj_id)
...     return obj.reference_cnt
...
```

回到示例中,看看有多少引用指向标记为 x 的列表。

```
>>> x = [1,2]
>>> idx = id(x)
>>> get_refs(idx)
1
```

当创建循环引用之后,观察计数器的变化。

```
>>> x.append(x)
>>> get_refs(idx)
2
>>> del x
>>> get_refs(idx)
1
```

可以看到,即使删除 x 后,计数器仍为 1。

这种情况可以强制使用垃圾收集器来处理。

```
>>> import gc
>>> gc.collect()
216
>>> get_refs(idx)  # 最终被移除
0
```

注意，上面的例子只是帮助你理解 Python 的内存管理。实际上，你不必手动启动垃圾收集器，因为 Python 可以决定何时高效地启动垃圾收集器。但是从垃圾收集的角度来看，在编程中最好使用元组而不是列表，因为元组是不可变的，不必像列表那样由垃圾收集器监控。

1.1.12 生成器

生成器提供及时高效的内存容器：

- 产生一个按需值流；
- 仅在使用 next( ) 函数时执行；
- yield( ) 函数产生一个值，但保存了函数的状态以备后用；
- 只能使用一次（即第一次使用后不能再次使用）。

```
>>> def generate_ints(N):
...     for i in range(N):
...         yield i  # yield()函数定义为生成器
...
>>> x=generate_ints(3)
>>> next(x)
0
>>> next(x)
1
>>> next(x)
2
>>> next(x)
Traceback (most recent call last):
  File "<stdin>", line 1, in <module>
StopIteration
```

清空生成器会引发 StopIteration 异常。在下一个块中会发生什么呢？

```
>>> next(generate_ints(3))
0
>>> next(generate_ints(3))
0
>>> next(generate_ints(3))
0
```

在上面的代码中，由于生成器并没有保存到一个变量中，因此生成器的状态无法存储。代码中的每一行实际都创建了一个新的生成器，因此每一行都是从头

开始迭代。当然,也可以直接对生成器进行迭代,示例如下:

```
>>> for i in generate_ints(5):  # 此处不需要赋值
...     print(i)
...
0
1
2
3
4
```

生成器能够维护一个内部状态,可以在 yield 之后返回到该状态。这也意味着生成器可以从上次离开的位置继续执行。示例如下:

```
>>> def foo():
...     print('hello')
...     yield 1
...     print('world')
...     yield 2
...
>>> x = foo()
>>> next(x)
hello
1

>>> # 此处安排一些其他任务
>>> next(x)  # 从离开的地方开始
world
2
```

生成器可以实现无限循环的算法,示例如下:

```
>>> def pi_series():  # 无穷级数收敛于 pi
...     sm = 0
...     i = 1.0; j = 1
...     while True:  # 一直循环
...         sm = sm + j/i
...         yield 4*sm
...         i = i + 2; j = j * -1
...
>>> x = pi_series()
>>> next(x)
4.0
>>> next(x)
2.666666666666667
>>> next(x)
3.466666666666667
>>> next(x)
2.8952380952380956
>>> next(x)
3.3396825396825403
>>> next(x)
2.9760461760461765
>>> next(x)
3.2837384837384844
>>> gen = (i for i in range(3))  # 列表
```

也可以使用 send( ) 向生成器的 yield 传递值,并恢复生成器的执行。

```
>>> def foo():
...     while True:
...         line=(yield)
...         print(line)
...
>>> x= foo()
>>> next(x)  # 开始
>>> x.send('I sent this to you')
I sent this to you
```

生成器也可以级联使用。

```
>>> def goo(target):
...     while True:
...         line=(yield)
...         target.send(line.upper()+'---')
...
>>> x= foo()
>>> y= goo(x)
>>> next(x)  # 开始
>>> next(y)  # 开始
>>> y.send('from goo to you')
FROM GOO TO YOU---
```

生成器也可以通过将列表的方括号改为圆括号来创建。

```
>>> x= (i for i in range(10))
>>> print(type(x))
<class 'generator'>
```

x 是一个生成器。itertools 模块是高效使用生成器的关键，可以克隆生成器，例如：

```
>>> x = (i for i in range(10))
>>> import itertools as it
>>> y, = it.tee(x,1)  # 克隆生成器
>>> next(y)  # 执行下一步
0
>>> list(zip(x,y))
[(1, 2), (3, 4), (5, 6), (7, 8)]
```

这种延迟执行的方法在处理大型数据集时特别有用。使用 map( ) 映射一个生成器。

```
>>> x = (i for i in range(10))
>>> y = map(lambda i:i**2,x)
>>> y
<map object at 0x7f939af36fd0>
```

注意，y 也是一个生成器，并且尚未进行任何计算。还可以使用 it.starmap 将函数映射到序列上。

生成器中的 yield from

迭代生成器的常用语法如下：

```
>>> def foo(x):
...     for i in x:
...         yield i
...
```

把生成器 x 输入 foo，

```
>>> x = (i**2 for i in range(3)) # 创建生成器
>>> list(foo(x))
[0, 1, 4]
```

使用 yield from，上面的代码可以用一行完成。

```
>>> def foo(x):
...     yield from x
...
```

然后，

```
>>> x = (i**2 for i in range(3)) # 重新创建生成器
>>> list(foo(x))
[0, 1, 4]
```

yield from 还可以做很多事情，假设我们用一个生成器来接收数据，示例如下：

```
>>> def accumulate():
...     sm = 0
...     while True:
...         n = yield # 从 send 接收
...         print(f'I got {n} in accumulate')
...         sm+=n
...
```

接下来看下如何运行，

```
>>> x = accumulate()
>>> x.send(None) # 启动
>>> x.send(1)
I got 1 in accumulate
>>> x.send(2)
I got 2 in accumulate
```

如果有一个函数组合，并希望将发送的值传递到嵌入式协同程序中呢？

```
>>> def wrapper(coroutine):
...     coroutine.send(None) # 启动
...     while True:
...         try:
...             x = yield            # 接收
...             coroutine.send(x)    # 传递
...         except StopIteration:
...             pass
...
```

接下来进行如下操作：

```
>>> w = wrapper(accumulate())
>>> w.send(None)
>>> w.send(1)
I got 1 in accumulate
>>> w.send(2)
I got 2 in accumulate
```

注意，这些传递的值直接传递给嵌套的协程。我们可以对其进行两次包装。

```
>>> w = wrapper(wrapper(accumulate()))
>>> w.send(None)
>>> w.send(1)
I got 1 in accumulate
>>> w.send(2)
I got 2 in accumulate
```

wrapper 函数可以缩写为

```
>>> def wrapper(coroutine):
...     yield from coroutine
...
```

此外，还能以同样的透明度自动处理嵌入的错误。下面是一个展平嵌套容器列表的例子：

```
>>> x = [1,[1,2],[1,[3]]]
>>> def flatten(seq):
...     for item in seq:
...         if hasattr(item,'__iter__'):
...             yield from flatten(item)
...         else:
...             yield item
...
>>> list(flatten(x))
[1, 1, 2, 1, 3]
```

生成器还有一种语法可以同时实现发送/接收操作，之前的 accumulate 函数有一个明显的问题，无法接收到累积的值。通过修改上一段代码中的一行代码，这个问题就可以解决。示例如下：

```
>>> def accumulate():
...     sm = 0
...     while True:
...         n = yield sm    # 从 send 接收, 发送 sm
...         print (f'I got {n} in accumulate and sm ={sm}')
...         sm+=n
...
```

接下来可以进行如下操作：

```
>>> x = accumulate()
>>> x.send(None) # 启动
0
```

注意，它返回了 0。

```
>>> x.send(1)
I got 1 in accumulate and sm =0
1
>>> x.send(2)
I got 2 in accumulate and sm =1
3
>>> x.send(3)
I got 3 in accumulate and sm =3
6
```

1.1.13 装饰器

装饰器（Decorators）是一种函数，它是用于拓展原来函数功能的一种函数。以下面的代码为例，由于 my_decorator 的输入 fn 是一个函数，而函数 fn 的输入参数未知，因此需要使用 args 和 kwargs 将它们传递到 my_decorator 的函数体内定义的 new_function 中。然后，在 new_function 中，我们只需调用输入函数 fn 并传入这些参数。在下面的代码中，最重要的一行是 return new_function，因为它通过返回函数体中的 new_function，实现了对 fn 函数的调用。因此，该装饰器函数既包含了 fn 函数的功能，又可以在 new_function 的主体中自由地执行其他任务。

```
>>> def my_decorator(fn):  # 函数作为输入
...     def new_function(*args,**kwargs):
...         print('this runs before function')
...         return fn(*args,**kwargs)  # 返回一个函数
...     return new_function
...
>>> def foo(x):
...     return 2*x
...
>>> goo = my_decorator(foo)
>>> foo(3)
6
>>> goo(3)
this runs before function
6
```

在上面的输出中，goo 函数忠实地再现了 foo 函数以及 new_function 函数中的输出。需要注意的是，在使用装饰器时，装饰器中构建的新函数应该与输入函数的功能逻辑正交。因为一旦输入函数与新函数发生混杂，会对函数的理解和调试带来很大的困难。下面的代码是合理使用装饰器的一个示例：假设我们想要监视一个函数的输入参数，可使用一个 log_arguments 装饰器，它既能实现输入参数的打印，又不会干扰该底层函数的业务逻辑。

```
>>> def log_arguments(fn):  # 函数作为输入
...     def new_function(*args,**kwargs):
...         print('positional arguments:')
...         print(args)
...         print('keyword arguments:')
...         print(kwargs)
...         return fn(*args,**kwargs)  # 返回一个函数
...     return new_function
...
```

你可以使用 @ 语法将装饰器叠加在函数定义上。这样做的好处是可以保留原始函数名，这意味着下游用户不必跟踪函数的另一个装饰版本。

```
>>> @log_arguments  # 这些也是可堆叠的
... def foo(x,y=20):
...     return x*y
...
>>> foo(1,y=3)
positional arguments:
(1,)
keyword arguments:
{'y': 3}
3
```

装饰器对于缓存也非常有用,它可以避免函数值的重复计算。

```
>>> def simple_cache(fn):
...     cache = {}
...     def new_fn(n):
...         if n in cache:
...             print('FOUND IN CACHE; RETURNING')
...             return cache[n]
...         # 否则,调用函数和记录值
...         val = fn(n)
...         cache[n] = val
...         return val
...     return new_fn
...
>>> def foo(x):
...     return 2*x
...
>>> goo = simple_cache(foo)
>>> [goo(i) for i in range(5)]
[0, 2, 4, 6, 8]
>>> [goo(i) for i in range(8)]
FOUND IN CACHE; RETURNING
FOUND IN CACHE; RETURNING
FOUND IN CACHE; RETURNING
FOUND IN CACHE; RETURNING
FOUND IN CACHE; RETURNING
[0, 2, 4, 6, 8, 10, 12, 14]
```

在上面的代码中,simple_cache 装饰器运行输入函数,然后将每个输出存储在 cache 字典中,键对应于函数输入。如果再次使用相同的输入调用该函数,则不会重新计算对应的函数值,而是从缓存中检索,如果输入函数需要长时间计算,则这样做非常高效。这种模式非常常见,现在已经包含在 Python 标准库的 functools.lru_cache 中。

> **编程技巧:装饰器模块**
>
> 一些 Python 模块作为装饰器分发(例如,用于创建命令行接口的单击模块),以便于在不更改源代码的情况下插入新功能。无论装饰器提供什么样的新功能,它都应该与被装饰的功能的业务逻辑不同。这样才能分离装饰器和装饰函数之间的关系。

装饰器对于在线程中执行某些函数也很有用。回想一下，线程是一组指令，CPU 可以与父进程分开运行。下面的代码就是通过装饰器来实现函数在单独的线程中运行的目的：

```
>>> def run_async(func):
...     from threading import Thread
...     from functools import wraps
...     @wraps(func)
...     def async_func(*args, **kwargs):
...         func_hl = Thread(target = func,
...                          args = args,
...                          kwargs = kwargs)
...         func_hl.start()
...         return func_hl
...     return async_func
...
```

functools 模块中的 wrap 函数可以修复函数签名。当你有一个小的副任务（比如通知）需要在主线程执行之外运行时，装饰器非常有用。下面是一个简单的函数来模拟一些假的任务：

```
>>> from time import sleep
>>> def sleepy(n=1,id=1):
...     print('item %d sleeping for %d seconds...'%(id,n))
...     sleep(n)
...     print('item %d done sleeping for %d seconds'%(id,n))
...
```

可参考以下代码块：

```
>>> sleepy(1,1)
item 1 sleeping for 1 seconds...
item 1 done sleeping for 1 seconds
>>> print('I am here!')
I am here!
>>> sleepy(2,2)
item 2 sleeping for 2 seconds...
item 2 done sleeping for 2 seconds
>>> print('I am now here!')
I am now here!
```

现在，使用装饰器来生成此函数的异步版本。

```
@run_async
def sleepy(n=1,id=1):
    print('item %d sleeping for %d seconds...'%(id,n))
    sleep(n)
    print('item %d done sleeping for %d seconds'%(id,n))
```

使用上述相同的语句块，可以获得以下打印输出序列：

```
sleepy(1,1)
print('I am here!')
sleepy(2,2)
print('I am now here!')

I am here!
item 1 sleeping for 1 seconds...
item 2 sleeping for 2 seconds
```

```
I am now here!
item 1 done sleeping for 1 seconds
item 2 done sleeping for 2 seconds
```

可以看到，在上面的代码中，最后一个 print 语句实际上是在单个函数完成之前执行的。这是因为程序执行的主线程处理这些打印语句，而独立线程的休眠时间不同。换句话说，在第一个示例中，最后的语句被前面的语句阻塞，必须等它们完成后才能执行最终的打印。而在第二个示例中使用装饰器实现独立线程时，主线程的语句没有阻塞，其他工作在单独的线程中进行。

装饰器的另一个常见用法是创建闭包。代码示例如下：

```
>>> def foo(a=1):
...     def goo(x):
...         return a*x # 使用外部的'a'
...     return goo
...
>>> foo(1)(10) # a=1
10
>>> foo(2)(10) # a=2
20
>>> foo(3)(10) # a=3
30
```

在上面的代码中，嵌套的 goo 函数需要一个参数 a，但在其函数签名中未定义。因此，foo 函数会生成嵌套的 goo 函数的不同版本，如示例所示，假设你有许多用户，每个用户都有不同的凭据来访问 goo 可以访问的数据。此时，foo 函数可以为 goo 函数封装这些凭据，使得每个用户都有其对应凭据的 goo 函数。这种方式简化了代码，因为 goo 的业务逻辑和函数签名无需改变，并且当 goo 超出作用域时，凭据会自动释放。

> **编程技巧：Python 线程**
>
> Python 中的线程主要用于使应用程序具有响应性。例如，如果你有一个包含许多按钮的 GUI，并且希望应用程序在单击每个按钮时对其做出反应，那么线程是合适的处理方式。再举一个例子，假设你正在从多个网站下载文件，那么使用单独的线程处理每个网站就非常有必要。
>
> 线程和进程之间的主要区别在于，进程有自己的划分资源。C 语言 Python（即 CPython）实现了一个全局解释器锁（GIL），防止线程争夺内部数据结构。因此，GIL 采用了过程粒度锁定机制，其中在任何时候只有一个线程可以访问资源。因此 GIL 简化了线程编程，因为同时运行多个线程需要复杂的簿记。GIL 的缺点是不能同时运行多个线程来加速计算受限的任务。注意，像

IronPython 这样的 Python 采用更精细的线程设计而不是 GIL 方法。

最后，在具有多核的现代系统上，多线程实际上可能会减慢速度。因为操作系统可能必须在不同的核之间切换线程，这会在线程切换机制中产生额外的开销，最终会减慢速度。CPython 在字节码级别实现 GIL，这意味着禁止跨不同线程的字节码指令同时执行。

1.1.14 迭代

迭代器允许对循环构造进行更精细的控制。

```
>>> a = range(3)
>>> hasattr(a,'__iter__')
True
>>> # 返回支持迭代的对象
>>> iter(a)
<range_iterator object at 0x7f939a6a6630>
>>> hasattr(_,'__iter__')
True
>>> for i in a: # 在循环中使用可迭代
...     print(i)
...
0
1
2
```

上面的 hasattr 是检查给定对象是否可迭代的传统方法。下面是一种新的方式。

```
>>> from collections.abc import Iterable
>>> isinstance(a,Iterable)
True
```

也可以使用 iter() 函数来创建哨兵（sentinel），这个概念在函数中非常有用。

```
>>> x=1
>>> def sentinel():
...     global x
...     x+=1
...     return x
...
>>> for k in iter(sentinel,5): #当 x=5 时停止
...     print(k)
...
2
3
4
>>> x
5
```

可以将其与文件对象 f 一起使用，例如 iter（f.readline，''），它会逐行读取文件中的内容，直到文件结束。

枚举

枚举意味着将符号名称与独特的常量值进行绑定。Python 的标准库中有一个 enum 模块可以支持枚举，示例如下：

```
>>> from enum import Enum
>>> class Codes(Enum):
...         START = 1
...         STOP = 2
...         ERROR = 3
...
>>> Codes.START
<Codes.START: 1>
>>> Codes.ERROR
<Codes.ERROR: 3>
```

一旦这些被定义，尝试更改它们将引发 AttributeError 异常。名称和值可以被提取出来。

```
>>> Codes.START.name
'START'
>>> Codes.START.value
1
```

枚举可以通过下面的方式迭代：

```
>>> [i for i in Codes]
[<Codes.START: 1>, <Codes.STOP: 2>, <Codes.ERROR: 3>]
```

可以使用合理的默认值直接创建枚举。

```
>>> Codes = Enum('Lookup','START,STOP,ERROR')
>>> Codes.START
<Lookup.START: 1>
>>> Codes.ERROR
<Lookup.ERROR: 3>
```

也可以查找与值对应的名称。

```
>>> Codes(1)
<Lookup.START: 1>
>>> Codes['START']
<Lookup.START: 1>
```

还可以使用 unique 装饰器来确保没有重复的值。

```
>>> from enum import unique
```

这将会引起 ValueError。

```
>>> @unique
... class UniqueCodes(Enum):
...     START = 1
...     END = 1 # 与 START 值相同
...
Traceback (most recent call last):
```

```
  File "<stdin>", line 2, in <module>
  File "/mnt/e/miniconda3/lib/python3.8/enum.py", line 860, in
  ↪   unique
    raise ValueError('duplicate values found in %r: %s' %
ValueError: duplicate values found in <enum 'UniqueCodes'>: END
↪    -> START
```

类型注释

Python 3 支持通过类型注释来为函数提供变量类型信息，以便其他工具（如 mypy）在分析大型代码库中检查类型冲突。类型注释不会影响 Python 的动态类型特性，这意味着除了创建单元测试之外，类型注释还提供了一种补充方式，通过揭示单元测试无法发现的缺陷来提高代码质量。

```
# filename: type_hinted_001.py

def foo(fname:str) -> str:
    return fname+'.txt'

foo(10) # 调用时使用了整数参数而不是字符串
```

在终端命令行上通过 mypy 运行此命令的方法是：

```
% mypy type_hinted_001.py
```

运行后产生以下错误：

```
type_hinted_001.py:6: error: Argument 1 to "foo" has incompatible
↪    type "int"; expected "str"
```

未进行类型注释的函数（即动态类型函数）可以与已注释的函数共存于同一个模块中。mypy 本身可能会尝试检测这些动态类型函数的类型错误，但这种行为是不稳定的。除了为函数提供类型注释，还可以直接提供默认值，如下所示：

```
>>> def foo(fname:str = 'some_default_filename') -> str:
...     return fname+'.txt'
...
```

类型不会从默认值的类型中推断得出。内置的 typing 模块同样提供了可用于类型提示的定义，如下所示：

```
>>> from typing import Iterable
>>> def foo(fname: Iterable[str]) -> str:
...     return "".join(fname)
...
```

在上面的代码中，声明了输入是一个可迭代的（例如，list）字符串，并且以该函数返回的单个字符串作为输出。Python 3.6 以后还支持对变量使用类型注释，如下所示：

```
>>> from typing import List
>>> x: str = 'foo'
>>> y: bool = True
>>> z: List[int] = [1]
```

请记住，这些添加在解释器中会被忽略，但在 mypy 中会被单独处理。类型注释也适用于类，如下所示：

```
>>> from typing import ClassVar
>>> class Foo:
...     count: ClassVar[int] = 4
...
```

因为类型提示会使代码变得复杂，特别是对于复杂的对象模式，所以类型注释可以分离到 pyi 文件中。然而，复杂性带来的负担可能会超过维护此类类型检查的好处。对于值得单独维护的关键代码，这种额外的负担可能是合理的。但除此之外，在开发和维护类型注释所需的工具进一步成熟之前，最好将其搁置一边。

Pathlib

Python 3 中引入了新的 pathlib 模块，使得在文件系统中的操作变得更加简单。在 pathlib 出现之前，你必须使用 os.walk 或其他一些目录搜索工具的组合来处理目录和路径。首先，我们从模块中导入 Path。

```
>>> from pathlib import Path
```

Path 对象本身可以提供非常有用的信息，例如：

```
>>> Path.cwd() # 获得当前目录
PosixPath('/mnt/d/class_notes_pub')
>>> Path.home() # 获得用户主目录
PosixPath('/home/unpingco')
```

你可以给对象一个起始路径，

```
>>> p = Path('./') # 指向当前目录
```

接下来，可以使用此方法搜索目录路径，

```
>>> p.rglob('*.log') # 搜索当前目录下具有 .log 扩展名的文件
<generator object Path.rglob at 0x7f939a5ecc80>
```

它将返回一个生成器，你可以遍历该生成器来获取目录中的所有 log 文件。每个迭代返回的元素都是一个 PosixPath 对象，具有自己的方法。

```
>>> item = list(p.rglob('*.log'))[0]
>>> item
PosixPath('altair.log')
```

例如，stat 方法可以提供文件的元数据。

```
>>> item.stat()
os.stat_result(st_mode=33279, st_ino=3659174697287897, st_dev=15,
↪    st_nlink=1, st_uid=1000, st_gid=1000, st_size=52602,
↪    st_atime=1608339360, st_mtime=1608339360,
↪    st_ctime=1608339360)
```

pathlib 模块还可以进行更多的操作，例如创建文件或处理单个路径的元素。

Asyncio

通过之前的学习，我们知道手动驱动生成器/协程可以实现交换和处理数据。实际上，可以将协程看作需要外部管理才能进行工作的对象。对于生成器来说，通常代码的业务逻辑会管理流程，但 async 可以简化这项工作并隐藏大规模处理时的复杂实现细节。

使用 async 关键字定义的函数。

```
>>> async def sleepy(n=1):
...     print(f'n = {n} seconds')
...     return n
...
```

我们可以像启动普通生成器一样启动它。

```
>>> x = sleepy(3)
>>> type(x)
<class 'coroutine'>
```

现在，我们用一个事件循环来驱动它。

```
>>> import asyncio
>>> loop = asyncio.get_event_loop()
```

接下来，loop 可以驱动协同程序。

```
>>> loop.run_until_complete(x)
n = 3 seconds
3
```

让我们把它放在一个同步阻塞循环中。

```
>>> from time import perf_counter
>>> tic = perf_counter()
>>> for i in range(5):
...     sleepy(0.1)
...
<coroutine object sleepy at 0x7f939a5f7ac0>
<coroutine object sleepy at 0x7f939a4027c0>
<coroutine object sleepy at 0x7f939a5f7ac0>
<coroutine object sleepy at 0x7f939a4027c0>
<coroutine object sleepy at 0x7f939a5f7ac0>
>>> print(f'elapsed time = {perf_counter()-tic}')
elapsed time = 0.0009255999993911246
```

什么都没发生！如果我们想在其他代码中使用 sleepy 函数，则需要使用 await 关键字。

```
>>> async def naptime(n=3):
...     for i in range(n):
...         print(await sleepy(i))
...
```

确切地说，await 关键字表示调用函数应该被挂起，直到 await 后面的目标完成，同时控制权被传递回事件循环。接下来，我们使用事件循环来驱动。

```
>>> loop = asyncio.get_event_loop()
>>> loop.run_until_complete(naptime(4))
n = 0 seconds
0
n = 1 seconds
1
n = 2 seconds
2
n = 3 seconds
3
```

注意，如果没有 await 关键字，naptime 函数将只返回 sleepy 对象，而不是这些对象的输出。而且函数体中必须包含异步代码，否则将会阻塞。目前，Python 模块正在开发中以支持这一框架（例如用于异步 Web 访问的 aiohttp），但这些模块仍在持续完善中。

让我们用另一个例子来展示事件循环如何为每个异步函数传递控制权。

```
>>> async def task1():
...     print('entering task 1')
...     await asyncio.sleep(0)
...     print('entering task 1 again')
...     print('exiting task 1')
...
>>> async def task2():
...     print('passed into task2 from task1')
...     await asyncio.sleep(0)
...     print('leaving task 2')
...
>>> async def main():
...     await asyncio.gather(task1(),task2())
...
```

所有这些都设置好后，我们用事件循环来启动。

```
>>> loop = asyncio.get_event_loop()
>>> loop.run_until_complete(main())
entering task 1
passed into task2 from task1
entering task 1 again
exiting task 1
leaving task 2
```

上面的代码中，await asyncio.sleep(0) 语句告诉事件循环将控制权传递给下一个项目，因为当前的项目将处于等待中，所以事件循环可以在此期间执行其他操作。这意味着任务可能会根据它们的执行时间以不同的顺序完成。代码示例如下：

```
>>> import random
>>> async def asynchronous_task(pid):
...     await asyncio.sleep(random.randint(0, 2)*0.01)
...     print('Task %s done' % pid)
...
>>> async def main():
...     await asyncio.gather(*[asynchronous_task(i) for i in
↪   range(5)])
...
```

```
>>> loop = asyncio.get_event_loop()
>>> loop.run_until_complete(main())
Task 0 done
Task 1 done
Task 2 done
Task 3 done
Task 4 done
```

这些异步执行操作在 asyncio 生态系统中得到了很好的实现。但对于没有进行这样配置的已有代码该怎么办呢？我们可以使用 concurrent.futures 将其整合到框架中，示例如下：

```
>>> from functools import wraps
>>> from time import sleep
>>> from concurrent.futures import ThreadPoolExecutor
>>> executor = ThreadPoolExecutor(5)
>>> def threadpool(f):
...     @wraps(f)
...     def wrap(*args, **kwargs):
...         return asyncio.wrap_future(executor.submit(f,
...                                                    *args,
...                                                    **kwargs))
...     return wrap
...
```

通过装饰一个阻塞的 sleepy 并使用 asyncio.wrap_future 将线程折叠到 asyncio 框架中。

```
>>> @threadpool
... def synchronous_task(pid):
...     sleep(random.randint(0, 1))
...     print('synchronous task %s done' % pid)
...
>>> async def main():
...     await asyncio.gather(synchronous_task(1),
...                          synchronous_task(2),
...                          synchronous_task(3),
...                          synchronous_task(4),
...                          synchronous_task(5))
...
```

运行事件循环，需注意输出结果是非顺序性的。

```
>>> loop = asyncio.get_event_loop()
>>> loop.run_until_complete(main())
synchronous task 3 done
synchronous task 5 done
synchronous task 1 done
synchronous task 2 done
synchronous task 4 done
```

调试和日记

调试 Python 最简单的方法是使用命令行。

```
% python -m pdb filename.py
```

这样你就会自动进入调试器。另外一种方法是在源代码文件中添加以下代码：

```
import pdb; pdb.set_trace()
```

这样可以在你希望设置断点的位置进行调试。另外，使用分号将代码放在一行可以更加方便地在编辑器中查找和删除。

另外，还可以通过以下方法在代码的任何地方提供一个完全交互式的 Python shell：

```
import code; code.interact(local=locals());
```

Python 3.7 以后新引入 breakpoint 函数，它与上面的 import pdb；pdb.set_trace（）行本质上是相同的，但它允许使用 PYTHONBREAKPOINT 环境变量来关闭或打开断点，如下所示：

```
% export PYTHONBREAKPOINT=0 python foo.py
```

因为环境变量被设置为零，foo.py 源代码中的断点行会被忽略。如果将环境变量设置为 1 则会产生相反的效果。除此之外，你还可以使用环境变量来选择调试器。例如，如果想用 IPython 调试器，则可以执行以下操作：

```
export PYTHONBREAKPOINT=ipdb.set_trace python foo.py
```

使用这个环境变量，你还可以在调用断点时运行自定义代码，例如：

```
# 文件名 break_code.py
def do_this_at_breakpoint():
    print ('I am here in do_this_at_breakpoint')
```

那么假设我们已经在 foo.py 文件中设置了断点，则可以执行以下操作：

```
% export PYTHONBREAKPOINT=break_code.do_this_at_breakpoint python
↪ foo.py
```

然后代码将会运行。请注意，由于这段代码没有调用调试器，因此执行不会在断点处停止。breakpoint 函数也可以接受参数。

```
breakpoint('a','b')
```

然后调用函数将会处理这些输入，如下所示：

```
# 文件名 break_code.py
def do_this_at_breakpoint(a,b):
    print ('I am here in do_this_at_breakpoint')
    print (f'argument a = {a}')
    print (f'argument b = {b}')
```

然后，在运行时将打印出这些变量值。请注意，你还可以在自定义的断点函数中通过 import pdb; pdb.set_trace（）来显式调用的调试器，它将使用内置的调试器来停止代码。

1.1.15 使用 Python 断言进行预调试

断言（assert）是为代码提供合理性检查的一种很好的方法。使用断言是增加代码可靠性的最快捷、最简单的方法。请注意，你可以在命令行上使用 -O 选项来关闭断言功能。

```
>>> import math
>>> def foo(x):
...     assert x>=0 # 输入条件
...     return math.sqrt(x)
...
>>> foo(-1)
Traceback (most recent call last):
  File "<stdin>", line 1, in <module>
  File "<stdin>", line 2, in foo
AssertionError
```

上述代码会抛出一个 AssertionError 异常。这是因为 foo 函数的输入条件没有被满足。

接下来看一个例子。

```
>>> def foo(x):
...     return x*2
...
```

任何能够理解乘法运算符的输入 x 都将通过，示例如下：

```
>>> foo('string')
'stringstring'
>>> foo([1,3,'a list'])
[1, 3, 'a list', 1, 3, 'a list']
```

上述代码的返回值可能不是你想要的结果。为了避免这种情况的发生，可以使用断言来强制要求输入为数字类型。

```
def foo(x):
    assert isinstance(x,(float,int,complex))
    return x*2
```

或者更好的方式是使用抽象数据类型。

```
import numbers
def foo(x):
    assert isinstance(x,numbers.Number)
    return x*2
```

关于在这种情况下是否应该使用 assert 是有争论的，因为 Python 的鸭子类型（duck typing）同样可以解决这个问题，但有时等待回溯报错并不适用于你的应用程序。断言的功能远远不止于输入类型检查，还可以用于其他方面。例如，假设你有一个问题，其中一个中间列表中所有项的总和应该等于 1。你可以在代码中使用 assert 来验证这一点，如果不成立，assert 会自动引发 AssertionError。因此，断言可以用于确保程序在运行时满足某些条件，从而提高代码的可靠性和可维护性。

1.1.16　使用 sys.settrace 进行堆栈追踪

Python 有一个强大的追踪功能，可以报告 Python 执行的每一行代码，并且还可以对这些行进行过滤。这种追踪功能对于调试复杂程序和性能分析非常有用。与使用调试器逐步执行程序相比，追踪功能可以提供更全面的代码执行信息，有助于定位 bug 和理解程序的执行流程。

以下是一个示例代码，它展示了如何在 Python 中使用追踪功能。

```
# filename:tracer_demo.py
# demo tracing in python

def foo(x=10,y=10):
    return x*y

def goo(x,y=10):
    y= foo(x,y)
    return x*y

if __name__ =="__main__":
    import sys
    def tracer(frame,event,arg):
        if event=='line': # 即将被执行的行
            filename, lineno = frame.f_code.co_filename,
                frame.f_lineno
            print(filename,end='\t')            # 文件名
            print(frame.f_code.co_name,end='\t')# 函数名
            print(lineno,end='\t')              # 文件中的行号
            print(frame.f_locals,end='\t')      # 局部变量
            argnames =
                frame.f_code.co_varnames[:frame.f_code.co_argcount]
            print(' arguments:',end='\t')
            print(str.join(', ',['%s:%r' % (i,frame.f_locals[i]) for
                i in argnames]))
        return tracer # 传递函数以备下次使用

    sys.settrace(tracer)
    foo(10,30)
    foo(20,30)
    goo(33)
```

关键的一步是将追踪函数传递给 sys.settrace，这将在堆栈帧上运行追踪函数并报告指定的元素。你还可以使用内置的跟踪器，例如：

```
% python -m trace --ignore-module=sys --trace filename.py >
    output.txt
```

这将产生大量的输出，因此你可能需要使用其他标志进行过滤。

> **编程技巧：Pysnooper**
>
> Pypi 上的 pynsooper 模块可以使得追踪变得非常容易实现和监控，强烈推荐使用。

1.1.17　使用 IPython 进行调试

你也可以在命令行中使用 IPython，例如：

```
% ipython --pdb <filename>
```

如果你想使用 IPython 调试器而不是默认调试器，也可以在源代码中这样做：

```
from IPython.core.debugger import Pdb
pdb=Pdb() # 创建实例
for i in range(10):
    pdb.set_trace() # 在这里设置断点
    print (i)
```

这提供了通常的 IPython 动态内省。这被分隔到了来自 PyPI 的 ipdb 包中。你也可以通过以下方式调用嵌入式 IPython shell：

```
import IPython
for i in range(10):
    if i > 7:
        IPython.embed() # 停止并嵌入 IPython shell
    print (i)
```

这种方法对于将 Python 嵌入 GUI 中时非常方便。在其他情况下，效果可能因人而异，但在紧急情况下这是一个不错的技巧。

1.1.18　从 Python 中进行日志记录

Python 有一个强大的内置日志记录包来实现日志记录，但需要进行一些设置，示例如下：

```
import logging, sys
logging.basicConfig(stream=sys.stdout, level=logging.INFO)
logging.debug("debug message") # 没有输出
logging.info("info message") # 输出
logging.error("error")   # 输出
```

注意，日志级别的数值是这样的：

```
>>> import logging
>>> logging.DEBUG
10
>>> logging.INFO
20
```

当日志记录级别设置 INFO 时，只有 INFO 及以上级别的消息会被记录。你还可以使用格式化器来格式化输出，示例如下：

```
import logging, sys
logging.basicConfig(stream=sys.stdout,
                    level=logging.INFO,
                    format="%(asctime)s - %(name)s -
                    ↪   %(levelname)s - %(message)s")

logging.debug("debug message") # 没有输出
logging.info("info message")
logging.error("error")
```

到目前为止，我们一直在使用根日志记录器，但是你可以根据日志记录器的名称创建多层次的有组织的日志记录。尝试在控制台中运行 demo_log1.py，看看会发生什么。

```
# 顶层程序

import logging, sys
from demo_log2 import foo, goo

log = logging.getLogger('main') # 设置日志记录器的名称
log.setLevel(logging.DEBUG)
handler = logging.StreamHandler(sys.stdout)
handler.setLevel(logging.DEBUG)

filehandler = logging.FileHandler('mylog.log')
formatter = logging.Formatter("%(asctime)s - %(name)s -
↪   %(funcName)s - %(levelname)s - %(message)s") # 设置格式

handler.setFormatter(formatter) # 设置格式
filehandler.setFormatter(formatter) # 设置格式
log.addHandler(handler) # 添加处理器以输出到控制台
log.addHandler(filehandler) # 添加处理器以输出到文件

def main(n=5):
    log.info('main called')
    [(foo(i),goo(i)) for i in range(n)]

if __name__ == '__main__':
    main()

# 附属于 demo_log1

import logging
log = logging.getLogger('main.demo_log2')

def foo(x):
    log.info('x=%r'%x)
    return 3*x
```

```
def goo(x):
    log = logging.getLogger('main.demo_log2.goo')
    log.info('x=%r'%x)
    log.debug('x=%r'%x)
    return 5*x**2

def hoo(x):
    log = logging.getLogger('main.demo_log2.hoo')
    log.info('x=%r'%x)
    return 5*x**2
```

现在，尝试更改 demo_log1.py 中主要日志记录器的名称，看看会发生什么。除非日志记录器是主日志记录器的下属，否则 demo_log2.py 中的日志记录不会被记录。你可以看到，将这段代码嵌入你的代码中可以轻松地启用各种代码诊断级别。你还可以为不同的日志级别设置多个处理器。

第 2 章

面向对象编程

Python 是一种面向对象的语言。面向对象编程有利于封装变量和函数，并将程序的各种关注点分离开来。这提高了可靠性，因为常见功能可以集中到相同的代码中。与 C++ 或 Java 相比，你不必编写自定义类来与 Python 的内置功能进行交互，因为面向对象编程并不是 Python 支持的唯一编程风格。事实上，我认为你希望将 Python 已经提供的内置对象和类连接在一起，而不是从头开始编写自己的类。在本节中，我们将建立起相关背景知识，这样你就可以根据需要编写自己的自定义类。

2.1 属性

对象封装的变量称为属性。在 Python 中，一切皆为对象。例如：

```
>>> f = lambda x:2*x
>>> f.x = 30 # 将属性附加到 x 函数对象上
>>> f.x
30
```

我们刚刚在函数对象上附加了一个属性。如果需要，甚至可以在函数内部引用它。

```
>>> f = lambda x:2*x*f.x
>>> f.x = 30 # 将属性附加到 x 函数对象上
>>> f.x
```

```
30
>>> f(3)
180
```

出于安全原因，对于某些内置对象（例如，__slots__），这种属性附加到对象上的方式会被阻止，但这不妨碍你理解这个概念。你可以使用 class 关键字来创建自己的对象，下面是最简单的自定义对象。

```
>>> class Foo:
...     pass
...
```

使用括号调用 Foo 对象来实例化它。

```
>>> f = Foo()   # 需要括号
>>> f.x = 30    # 添加属性
>>> f.x
30
```

除了上面使用内置函数对象逐个附加属性的方法，还可以使用 __init__ 方法对该类的所有实例进行操作。

```
>>> class Foo:
...     def __init__(self):  # 注意双下划线
...         self.x = 30
...
```

self 关键字引用了所创建的实例。我们可以像下面这样实例化这个对象。

```
>>> f = Foo()
>>> f.x
30
```

__init__ 构造函数将属性附加到所有这样创建的对象中。

```
>>> g = Foo()
>>> g.x
30
```

你还可以向 __init__ 函数提供在实例化时调用的参数。

```
>>> class Foo:
...     def __init__(self,x=30):
...         self.x = x
...
>>> f = Foo(99)
>>> f.x
99
```

注意，__init__ 是一个 Python 函数，它遵循 Python 函数的语法。__init__ 的双下划线表示该函数具有特殊的低等级状态。

> **编程技巧：私有属性与公有属性**
>
> 与许多面向对象的语言不同，Python 不需要在语言中说明私有属性或公有属性，而是通过约定进行管理。例如，以单个下划线开头的属性（仅根据约定）被视为私有属性，尽管语言本身中没有为它们提供特殊的状态。

2.2 方法

方法是附加到对象的函数，并可以访问内部对象属性。它们在类的主体中定义。

```
>>> class Foo:
...     def __init__(self,x=30):
...         self.x = x
...     def foo(self,y=30):
...         return self.x*y
...
>>> f = Foo(9)
>>> f.foo(10)
90
```

注意，可以使用 self 变量在 foo 函数体内部访问附加在 self.x 上的变量。在 Python 编程中，一种常见的做法是将所有不变的变量打包在 __init__ 函数的属性中，并配置方法将经常变化的变量设为函数变量供用户调用。此外，self 可以在方法调用之间维护状态，以便对象可以维护内部历史记录并更改对象方法的相应行为。

注意，方法至少有一个参数（即 self）。可参见以下错误：

```
>>> f.foo(10)     # 正常
90
>>> f.foo(10,3) # 出现错误
Traceback (most recent call last):
  File "<stdin>", line 1, in <module>
TypeError: foo() takes from 1 to 2 positional arguments but 3 were given
```

在上面的代码中，Python 将 f.foo(10) 视为 Foo.foo(f, 10)，因此第一个参数是实例 f，在方法定义中将其称为 self。因此，从 Python 的角度来看，已有两个参数。

> **编程技巧：函数和方法**
>
> 访问对象的属性是方法和函数之间的区别。例如，我们可以创建一个函数并将其附加到现有对象上，如下所示：
> ```
> >>> f = Foo()
> >>> f.func = lambda i:i*2
> ```
> 这是一个合法的函数，可以像方法 f.func(10) 一样调用它，但该函数无法访问 f 的任何内部属性，必须从调用中获取其所有参数。

方法可以调用同一对象中的其他方法，只要在调用时使用 self 前缀。像加法（"+"运算符）之类的操作也可以被定义为方法。

```
>>> class Foo:
...     def __init__(self,x=10):
...         self.x = x
...     def __add__(self,y): # 重载"+"运算符
...         return self.x + y.x
...
>>> a=Foo(x=20)
>>> b=Foo()
>>> a+b
30
```

> **编程技巧：将对象作为函数调用**
>
> 在 Python 中，函数也是对象，通过在类中添加一个 __call__ 方法，就可以使类像函数一样可调用。例如：
>
> ```
> >>> class Foo:
> ...     def __call__(self,x):
> ...         return x*10
> ...
> ```
>
> 接下来进行如下操作：
>
> ```
> >>> f = Foo()
> >>> f(10)
> 100
> ```
>
> 这种方法的优点在于可以在 __init__ 函数中提供额外的变量，然后就可以像使用任何其他函数一样使用这个对象。

2.3 继承

继承有助于代码重用。示例如下：

```
>>> class Foo:
...     def __init__(self,x=10):
...         self.x = x
...     def compute_this(self,y=20):
...         return self.x*y
...
```

假设我们想要有一个新类，新类中 compute_this 的工作方法需要更改，那么我们不需要重写类，只需继承它并更改（即覆盖）我们不喜欢的部分即可。

```
>>> class Goo(Foo): # 继承自 Foo
...     def compute_this(self,y=20):
...         return self.x*y*1000
...
```

现在，我们更新了 Foo 中的 compute_this 函数。

```
>>> g = Goo()
>>> g.compute_this(20)
200000
```

这个想法是通过继承来重用自己的代码（或者更好的是，重用他人的代码）。Python 还支持多重继承和代理（通过 super 关键字）。

假设我们想要实现一个特殊的 __repr__ 函数，可以从内置 list 对象进行继承。

```
>>> class MyList(list): # 从内置 list 对象继承
...     def __repr__(self):
...         list_string = list.__repr__(self)
...         return list_string.replace(' ','')
...
>>> MyList([1,3]) # 输出中没有空格
[1,3]
>>> list([1,3]) # 输出中有空格
[1, 3]
```

> **编程技巧：一个好的 __repr__ 的优点**
>
> repr 内置函数触发 __repr__ 方法，该方法将对象表示为字符串。严格来说，repr 应该返回一个字符串，当使用内置的 eval() 函数进行评估时，返回给定对象的实例。实际上，repr 返回一个字符串表示对象，这是一个很好的为对象添加标记的机会，使对象更易于在交互式解释器或 debugger 中进行理解。下面是一个示例：
>
> ```
> >>> class I:
> ...     def __init__(self,left,right,isopen=True):
> ...         self.left, self.right = left, right # 区间边缘
> ...         self.isopen = isopen
> ...     def __repr__(self):
> ...         if self.isopen:
> ...             return '(%d,%d)'%(self.left,self.right)
> ...         else:
> ...             return '[%d,%d]'%(self.left,self.right)
> ...
> >>> a = I(1,3) # 开区间
> >>> a
> (1,3)
> >>> b = I(11,13,False) # 闭区间
> >>> b
> [11,13]
> ```
>
> 在这个例子中，用括号或方括号表示后可以清楚地看出给定的区间是开的还是闭的。这类提示将使理解这些对象变得更加容易。

一旦你可以编写自己的类，就可以重现其他 Python 对象的行为，例如，迭代对象：

```
>>> class Foo:
...     def __init__(self,size=10):
...         self.size = size
...     def __iter__(self): # 产生迭代
...         self.counter = list(range(self.size))
...         return self # 返回具有 next( )方法的对象
...     def __next__(self): # 进行迭代
...         if self.counter:
...             return self.counter.pop()
...         else:
...             raise StopIteration
...
>>> f = Foo()
>>> list(f)
[9, 8, 7, 6, 5, 4, 3, 2, 1, 0]
>>> for i in Foo(3): # 迭代结束
...     print(i)
...
2
1
0
```

2.4 类变量

到目前为止，我们一直在讨论对象实例的属性和方法。其实除此之外，你可以使用类变量将变量直接绑定到类上。

```
>>> class Foo:
...     class_variable = 10 # 这里定义的变量与类相关,而不是特定的实例
...
>>> f = Foo()
>>> g = Foo()
>>> f.class_variable
10
>>> g.class_variable
10
>>> f.class_variable = 20
>>> f.class_variable # 修改实例的变量,只会改变该实例的属性,而不会影响其他实例或者类
20
>>> g.class_variable # 无修改
10
>>> Foo.class_variable   # 无修改
10
>>> Foo.class_variable  = 100 # 对类变量进行修改
>>> h = Foo()
>>> f.class_variable # 无修改
20
>>> g.class_variable # 即使已存在,也在此处修改
100
>>> h.class_variable # 此处修改
100
```

这不仅适用于变量，也适用于函数，但只适用于使用 @classmethod 装饰器

的函数。请注意，类变量的存在并不会自动被类定义中的其他部分所知。例如：

```
>>> class Foo:
...     x = 10
...     def __init__(self):
...         self.fx = x**2 # x 未知
...
```

由于类变量 x 对于 __init__ 属于未知，因此会产生以下错误提示：NameError: global name 'x' is not defined。这个错误可以通过提供对 x 的完整类引用来解决，如下所示：

```
>>> class Foo:
...     x = 10
...     def __init__(self):
...         self.fx = Foo.x**2 # 完整引用 x
...
```

不过，最好避免将类名硬编码到代码中，这会使下游继承变得脆弱。

2.5 类函数

函数可以使用 classmethod 装饰器将函数附加到类中成为类函数，例如：

```
>>> class Foo:
...   @classmethod
...   def class_function(cls,x=10):
...       return x*10
...
>>> f = Foo()
>>> f.class_function(20)
200
>>> Foo.class_function(20) # 不需要实例
200
>>> class Foo:
...     class_variable = 10
...     @classmethod
...     def class_function(cls,x=10):
...         return x*cls.class_variable # 使用类变量
...
>>> Foo.class_function(20) # 不需要实例,使用类变量
200
```

这对于将参数传输给所有类实例非常有用，例如：

```
>>> class Foo:
...     x=10
...     @classmethod
...     def foo(cls):
...         return cls.x**2
...
>>> f = Foo()
```

```
>>> f.foo()
100
>>> g = Foo()
>>> g.foo()
100
>>> Foo.x = 100  # 改变类变量
>>> f.foo()  # 实例也随之改变
10000
>>> g.foo()  # 实例也随之改变
10000
```

由于类本身拥有类变量，因此这种变化的跟踪可能比较棘手，例如：

```
>>> class Foo:
...     class_list = []
...     @classmethod
...     def append_one(cls):
...         cls.class_list.append(1)
...
>>> f = Foo()
>>> f.class_list
[]
>>> f.append_one()
>>> f.append_one()
>>> f.append_one()
>>> g = Foo()
>>> g.class_list
[1, 1, 1]
```

注意，新实例 g 获得了通过 f 实例对类变量所做的更改。接下来，如果进行如下操作：

```
>>> del f,g
>>> Foo.class_list
[1, 1, 1]
```

由于类变量是附加在类定义上的，因此删除类实例并不会影响类变量。当你以这种方式设置类变量时，请确保这是预期的行为。

类变量和类方法是不惰性求值的。当我们讨论到数据类（dataclasses）时，这一点将会更加明显。例如：

```
>>> class Foo:
...     print('I am here!')
...
I am here!
```

注意，我们没有为了执行 print 语句而实例化这个类。这对于面向对象的设计有着重要的微妙意义。有时，平台特定的参数会作为类变量插入，这样它们在类的任何实例被实例化时就会被设置好，例如：

```
>>> class Foo:
...     _setup_const = 10  # 平台特定的信息
...     def some_function(self,x=_setup_const):
...         return 2*x
...
```

```
>>> f = Foo()
>>> f.some_function()
20
```

2.6 静态方法

与类方法不同,staticmethod 附加到类的定义但不需要访问内部变量。示例如下:

```
>>> class Foo:
...     @staticmethod
...     def mystatic(x,y):
...         return x*y
...
```

staticmethod 不像普通实例化方法或 classmethod 那样需要访问内部的 self 或 cls。它只是一种将函数附加到类的方法。因此,

```
>>> f = Foo()
>>> f.mystatic(1,3)
3
```

2.7 哈希对子变量隐藏父变量

按照惯例,以单下划线字符开头的方法和属性被认为是私有的,以双下划线开头的方法和属性在内部会以类名进行哈希处理。

```
>>> class Foo:
...     def __init__(self):
...         self.__x=10
...     def count(self):
...         return self.__x*30
...
```

注意,count 函数使用了双下划线变量 self.__x。

```
>>> class Goo(Foo):  # 子类具有自己的 .__x 属性
...     def __init__(self,x):
...         self.__x=x
...
>>> g=Goo(11)
>>> g.count()  # 报错
Traceback (most recent call last):
  File "<stdin>", line 1, in <module>
  File "<stdin>", line 5, in count
AttributeError: 'Goo' object has no attribute '_Foo__x'
```

这意味着 Foo 声明中的 __x 变量与 Foo 类相关联。这是为了防止潜在的子类用 Foo.count() 函数时使用子类的变量(例如 self._x,不带双下划线)。

2.8 委托函数

在下面的继承链中，尝试思考 abs 函数是从哪里派生的。

```
>>> class Foo:
...     x = 10
...     def abs(self):
...         return abs(self.x)
...
>>> class Goo(Foo):
...     def abs(self):
...         return abs(self.x)*2
...
>>> class Moo(Goo):
...     pass
...
>>> m = Moo()
>>> m.abs()
20
```

当 Python 遇到 m.abs() 时，首先会检查 Moo 类是否实现了 abs() 函数。如果没有实现，那么 Python 会从左到右在继承链中查找，从而找到 Goo。如果 Goo 实现了 abs() 函数，则会使用这个函数，但它需要 self.x，而这个 self. 是 Goo 从其父类 Foo 那里获取的，用于完成计算。Goo 中的 abs() 函数依赖于内置的 Python abs() 函数。

2.9 使用 super 进行委托

Python 的 super 方法是一种按照类的方法解析顺序（Method Resolution Order，MRO）运行函数的一种方式。对于 super 来说，更好的名称应该是 MRO 中的下一个方法。在下面的例子中，类 A 和类 B 都继承自 Base。

```
>>> class Base:
...     def abs(self):
...         print('in Base')
...         return 1
...
>>> class A(Base):
...     def abs(self):
...         print('in A.abs')
...         oldval=super(A,self).abs()
...         return abs(self.x)+oldval
...
```

```
>>> class B(Base):
...     def abs(self):
...         print('in B.abs')
...         oldval=super(B,self).abs()
...         return oldval*2
...
```

设置好后，用图 2.1 中的继承树创建一个继承了 A 和 B 的新类。

```
>>> class C(A,B):
...     x=10
...     def abs(self):
...         return super(C,self).abs()
...
>>> c=C()  # 创建类 C 实例
>>> c.abs()
in A.abs
in B.abs
in Base
12
```

发生了什么？如图 2.1 所示，方法解析首先在类 C 中查找 abs 函数，找到后，会按照继承顺序依次执行。它会先找到类 A 中的 abs 函数并执行，然后继续前进到 MRO 中的下一个类（类 B），在那里也找到 abs 函数并执行。因此，super 方法基本上是将这些函数串联在一起，形成一个链式调用的过程。这种方式确保了在多重继承中的方法调用顺序是按照继承链的顺序进行的。

现在让我们看看，当改变方法解析顺序，从类 A 开始时会发生什么，如下所示：

```
>>> class C(A,B):
...     x=10
...     def abs(self):
...         return super(A,self).abs()
...
>>> c=C()  # 创建类 C 实例
>>> c.abs()
in B.abs
in Base
2
```

注意，它在类 A 之后执行方法解析（见图 2.2）。我们可以改变继承的顺序，看看这会如何影响 super 的解析顺序。

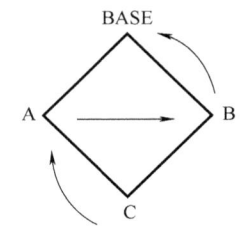

图 2.1　方法解析顺序

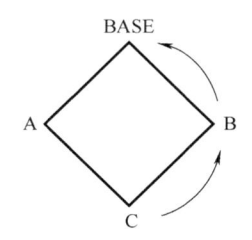

图 2.2　方法解析顺序

```
>>> class C(B,A):  # 改变MRO
...     x=10
...     def abs(self):
...         return super(B,self).abs()
...
>>> c=C()
>>> c.abs()
in A.abs
in Base
11
>>> class C(B,A):  # 同样的MRO,不同的super
...     x=10
...     def abs(self):
...         return super(C,self).abs()
...
>>> c=C()
>>> c.abs()
in B.abs
in A.abs
in Base
22
```

总之，super 允许你混合和匹配对象，根据对特定方法的解析方式，产生不同的使用方法。这为代码增加了另一个自由度，但也增加了另一层复杂性。

2.10 元编程：猴子补丁

猴子补丁（Monkey Patching）意味着在正式的类定义之外向类或实例添加方法。这样做会使得调试变得非常困难，因为你无法确切地知道函数被劫持的位置。虽然极不建议这样做，但也可以对类方法进行 Monkey Patching：

```
>>> import types
>>> class Foo:
...     class_variable = 10
...
>>> def my_class_function(cls,x=10):
...     return x*10
...
>>> Foo.my_class_function = types.MethodType(my_class_function,
↪   Foo)
>>> Foo.my_class_function(10)
100
>>> f = Foo()
>>> f.my_class_function(100)
1000
```

这种方法可以用于临时调试。

```
>>> import types
>>> class Foo:
...     @classmethod
```

```
...     def class_function(cls,x=10):
...         return x*10
...
>>> def reported_class_function():
...     # 将原始函数隐藏在闭包中
...     orig_function = Foo.class_function
...     # 定义替换函数
...     def new_class_function(cls,x=10):
...         print('x=%r' % x)
...         return orig_function(x)# 使用原始函数返回
...     return new_class_function # 返回一个函数
...
>>> Foo.class_function(10) # 原始方法
100
>>> Foo.class_function = types.MethodType(reported_class_
function(), Foo)
>>> Foo.class_function(10) # 新方法
x=10
100
```

> **编程技巧：保持简单**
>
> 在 Python 中，很容易对类进行花哨的操作，但最好的做法是在概念上简化和统一代码、避免代码冗余，并保持组织上的简单性时使用类和面向对象编程。要了解 Python 面向对象设计的完美应用，请研究 Networkx[2] 图模块。

2.11 抽象基类

collections 模块包含抽象基类，其具有两个主要功能。首先，它们提供了一种检查给定自定义对象是否具有所需接口的方式，可以使用 isinstance 或 issubclass。其次，它们为新对象提供了满足特定软件模式的最小要求。让我们考虑一个函数，比如 g = lambda x：x**2。假设想要测试 g 是否是可调用的，一种检查的方法是使用 callable 函数，如下所示：

```
>>> g = lambda x:x**2
>>> callable(g)
True
```

另一种方法是使用抽象基类，如下所示：

```
>>> from collections.abc import Callable
>>> isinstance(g,Callable)
True
```

抽象基类将这种能力扩展到给定的接口，正如 Python 主文档中所描述的那样。例如，要检查自定义对象是否可迭代，可以这样做：

```
>>> from collections.abc import Iterable
>>> isinstance(g,Iterable)
False
```

除了这种接口检查之外，抽象基类还允许对象设计者指定最小方法集，并通过继承获取特定抽象基类的其他方法。例如，如果我们想编写类似字典的对象，那么可以继承 MutableMapping 抽象基类，然后编写 __getitem__、__setitem__、__delitem__、__iter__、__len__ 等方法。这样，我们就可以通过继承获得其他 MutableMapping 方法，比如 clear()、update() 等。

ABCMeta 编程

抽象基类还可以通过使用 abc.abstractmethod 装饰器来强制子类方法的实现。例如：

```
>>> import abc
>>> class Dog(metaclass=abc.ABCMeta):
...     @abc.abstractmethod
...     def bark(self):
...         pass
...
```

这意味着 Dog 的所有子类都必须实现一个 bark 方法，否则将抛出 TypeError。装饰器将该方法标记为 abstract。

```
>>> class Pug(Dog):
...     pass
...
>>> p = Pug() # 抛出 TypeError
Traceback (most recent call last):
  File "<stdin>", line 1, in <module>
TypeError: Can't instantiate abstract class Pug with abstract
↪ methods bark
```

所以必须在子类中实现 bark 方法。

```
>>> class Pug(Dog):
...     def bark(self):
...         print('Yap!')
...
>>> p = Pug()
```

接下来，

```
>>> p.bark()
Yap!
```

除了从基类进行子类化，还可以使用 register 方法将另一个类变为子类，前提是它实现了所需的抽象方法，如下所示：

```
>>> class Bulldog:
...     def bark(self):
...         print('Bulldog!')
...
>>> Dog.register(Bulldog)
<class '__console__.Bulldog'>
```

这样，即使 Bulldog 不是作为 Dog 的子类编写的，它仍然被子类化：

```
>>> issubclass(Bulldog, Dog)
True
>>> isinstance(Bulldog(), Dog)
True
```

尽管抽象方法的具体实现是子类编写者的责任，但你仍然可以使用 super 来运行父类中的主定义。例如：

```
>>> class Dog(metaclass=abc.ABCMeta):
...     @abc.abstractmethod
...     def bark(self):
...         print('Dog bark!')
...
>>> class Pug(Dog):
...     def bark(self):
...         print('Yap!')
...         super(Pug,self).bark()
...
```

接下来，

```
>>> p= Pug()
>>> p.bark()
Yap!
Dog bark!
```

2.12 描述符

描述符将 Python 对象创建技术中的内部抽象暴露出来，以供更广泛的使用。描述符最简单的使用方式是将输入验证隐藏在对象中，使用户无需关注。例如：

```
>>> class Foo:
...     def __init__(self,x):
...         self.x = x
...
```

这个对象有一个名为 x 的简单属性。用户可以随意执行以下操作：

```
>>> f = Foo(10)
>>> f.x = 999
>>> f.x = 'some string'
>>> f.x = [1,3,'some list']
```

换句话说，属性 x 可以被分配给这些不同的类型。这可能不是你想要的结果。如果你想确保该属性只能被分配给整数，那么需要一种方法来强制执行。这就是描述符发挥作用的地方。具体操作方法如下：

```
>>> class Foo:
...     def __init__(self,x):
```

```
...         assert isinstance(x,int) # 强制要求
...         self._x = x
...     @property
...     def x(self):
...         return self._x
...     @x.setter
...     def x(self,value):
...         assert isinstance(value,int) # 强制要求
...         self._x = value
...
```

上面的代码中，有趣的部分发生在 property 装饰器上。这是我们明确改变 Python 处理 x 属性方式的地方。我们不再需要直接访问 Foo.__dict__ 来设置该属性的值，而是可以调用 x.setter 函数来直接处理赋值。同样，这也适用于获取属性的值，如下所示使用了 x.getter 装饰器：

```
>>> class Foo:
...     def __init__(self,x):
...         assert isinstance(x,int) # 强制要求
...         self._x = x
...     @property
...     def x(self):
...         return self._x
...     @x.setter
...     def x(self,value):
...         assert isinstance(value,int) # 强制要求
...         self._x = value
...     @x.getter
...     def x(self):
...         print('using getter!')
...         return self._x
...
```

使用 @getter 的好处是可以在每次访问属性时返回一个计算值。例如，我们可以执行以下操作：

```
@x.getter
def x(self):
    return self._x * 30
```

现在可以在用户不知道的情况下动态计算属性了。关键的一点是：整个描述符机制可以从类定义中抽象出来。当在同一个类定义中多次重复使用相同的描述符集合时，逐个为每个属性重写将会容易出错且繁琐，这时这种抽象就非常有用。例如，假设我们有一个 FloatDescriptor 类和另一个描述 Car 的类：

```
>>> class FloatDescriptor:
...     def __init__(self):
...         self.data = dict()
...     def __get__(self, instance, owner):
...         return self.data[instance]
...     def __set__(self, instance, value):
...         assert isinstance(value,float)
...         self.data[instance] = value
...
```

```
>>> class Car:
...     speed = FloatDescriptor()
...     weight = FloatDescriptor()
...     def __init__(self,speed,weight):
...         self.speed = speed
...         self.weight = weight
...
```

注意，FloatDescriptor 在 Car 的类定义中显示为类变量。这将在之后产生重要的影响，但现在让我们看看这是如何工作的。

```
>>> f = Car(1,2)  # 触发 AssertionError
Traceback (most recent call last):
  File "<stdin>", line 1, in <module>
  File "<stdin>", line 5, in __init__
  File "<stdin>", line 7, in __set__
AssertionError
```

触发 AssertionError 是因为描述符要求参数为浮点。

```
>>> f = Car(1.0,2.3)  # 因为满足了 FloatDescriptor 的要求,因此无 AssertionError
>>> f.speed = 10      # 触发 AssertionError
Traceback (most recent call last):
  File "<stdin>", line 1, in <module>
  File "<stdin>", line 7, in __set__
AssertionError
>>> f.speed = 10.0    # 无 AssertionError
```

上面的代码中，我们使用 FloatDescriptor 将 Car 类的属性管理和验证抽象出来，然后可以在其他类中重用 FloatDescriptor。然而，这里有一个重要的注意事项，因为我们必须在类级别使用 FloatDescriptor，以便正确地连接描述符。这意味着必须确保将实例属性的赋值放在正确的实例上。这就是为什么 FloatDescriptor 构造函数中的 self.data 是一个字典。我们使用实例本身作为该字典的键，以确保属性被放置在正确的实例上，如下所示：self.data [instance] = value。对于不可哈希的类，这种方法可能会失败，因为它们无法用作字典的键。__get__ 方法具有 owner 参数是因为这些问题可以通过元类来解决，但这超出了我们的范围。

总之，描述符是 Python 类在内部使用的低级机制，用于管理方法和属性。它们还提供了一种将类属性的管理抽象为独立的描述符类的方式，这些描述符类可以在多个类之间共享。对于不可哈希的类，描述符可能会变得棘手，并且在超出我们讨论范围的情况下，扩展此模式可能会遇到其他问题。《Python Essential Reference》[1]这本书是学习高级 Python 的最佳参考书之一。

2.13 具名元组和数据类

具名元组允许更容易、可读性更强地访问元组。例如:
```
>>> from collections import namedtuple
>>> Produce = namedtuple('Produce','color shape weight')
```
这里创建了一个名为 Produce 的新类,该类具有属性 color、shape 和 weight。注意,属性规范中不能有 Python 关键字或重复的名称。要使用这个新类,只需像其他类一样实例化它,
```
>>> mango = Produce(color='g',shape='oval',weight=1)
>>> print (mango)
Produce(color='g', shape='oval', weight=1)
```
注意,我们可以使用常用的索引获取元组的元素。
```
>>> mango[0]
'g'
>>> mango[1]
'oval'
>>> mango[2]
1
```
我们可以通过使用已命名的属性来得到相同的结果。
```
>>> mango.color
'g'
>>> mango.shape
'oval'
>>> mango.weight
1
```
元组解包与常规元组一样。
```
>>> i,j,k = mango
>>> i
'g'
>>> j
'oval'
>>> k
1
```
你还可以获得属性的名称。
```
>>> mango._fields
('color', 'shape', 'weight')
```
我们还可以通过使用 _replace 方法替换现有属性的值来创建新的具名元组对象,如下所示:
```
>>> mango._replace(color='r')
Produce(color='r', shape='oval', weight=1)
```

在底层，具名元组会自动生成代码来实现对应的类（在此例中为 Produce）。这种自动生成代码以实现特定类的理念在 Python 3.7+ 中得到了扩展，并被用于数据类。

数据类

在 Python 3.7+ 中，dataclasses 将代码生成的概念从具名元组扩展到更通用的类似数据的对象。

```
>>> from dataclasses import dataclass
>>> @dataclass
... class Produce:
...     color: str
...     shape: str
...     weight: float
...
>>> p = Produce('apple','round',2.3)
>>> p
Produce(color='apple', shape='round', weight=2.3)
```

注意，这里的类型不是强制的。

```
>>> p = Produce(1,2,3)
>>> p
Produce(color=1, shape=2, weight=3)
```

使用 dataclass 装饰器，可以获得许多额外的方法。

```
>>> dir(Produce)
['__annotations__', '__class__', '__dataclass_fields__',
'__dataclass_params__', '__delattr__', '__dict__', '__dir__',
'__doc__', '__eq__', '__format__', '__ge__', '__getattribute__',
'__gt__', '__hash__', '__init__', '__init_subclass__', '__le__',
'__lt__', '__module__', '__ne__', '__new__', '__reduce__',
'__reduce_ex__', '__repr__', '__setattr__', '__sizeof__',
'__str__', '__subclasshook__', '__weakref__']
```

其中，__hash__() 和 __eq__() 可作为字典的 keys 使用，但必须使用 frozen=True 关键字参数，如下所示：

```
>>> @dataclass(frozen=True)
... class Produce:
...     color: str
...     shape: str
...     weight: float
...
>>> p = Produce('apple','round',2.3)
>>> d = {p: 10} # 实例作为 key
>>> d
{Produce(color='apple', shape='round', weight=2.3): 10}
```

如果希望类根据输入元组排序，也可以使用 order=True。默认值可以按以下方式分配：

```
>>> @dataclass
... class Produce:
```

```
...     color: str = 'red'
...     shape: str = 'round'
...     weight: float = 1.0
...
```

与具名元组不同,你可以使用自定义方法。

```
>>> @dataclass
... class Produce:
...     color  : str
...     shape  : str
...     weight : float
...     def price(self):
...         return  0 if self.color=='green' else self.weight*10
...
```

与具名元组不同,数据类(dataclass)是不可迭代的。dataclass 可以使用辅助函数。field 函数允许你指定如何默认创建某些声明的属性。下面的示例使用一个 list factory 来避免类的所有实例共享同一类变量 list:

```
>>> from dataclasses import field
>>> @dataclass
... class Produce:
...     color  : str = 'green'
...     shape  : str = 'flat'
...     weight : float = 1.0
...     track  : list = field(default_factory=list)
...
```

Produce 的两个不同实例具有不同的可变 track 列表,这避免了在初始值设定项中使用可变对象的问题。dataclass 的其他参数允许自动定义对象的顺序或使其不可变,如下所示:

```
>>> @dataclass(order=True,frozen=True)
... class Coor:
...     x: float = 0
...     y: float = 0
...
>>> c = Coor(1,2)
>>> d = Coor(2,3)
>>> c < d
True
>>> c.x = 10
Traceback (most recent call last):
  File "<stdin>", line 1, in <module>
  File "<string>", line 4, in __setattr__
dataclasses.FrozenInstanceError: cannot assign to field 'x'
```

asdict 函数可以轻松地将数据类转换为常规 Python 字典,这对于序列化非常有用。注意,这只会转换实例的属性。

```
>>> from dataclasses import asdict
>>> asdict(c)
{'x': 1, 'y': 2}
```

如果你有依赖于其他初始化变量的变量,但不想在每个新实例中都自动创建它们，则可以使用 field 函数，如下所示：

```
>>> @dataclass
... class Coor:
...     x : float = 0
...     y : float = field(init=False)
...
>>> c = Coor(1) # y 没有初始化
>>> c.y = 2*c.x # y 初始化
>>> c
Coor(x=1, y=2)
```

更简单的方法是使用 __post_init__ 方法。注意：__init__ 方法是由 dataclass 自动生成的。

```
>>> @dataclass
... class Coor:
...     x : float = 0
...     y : float = field(init=False)
...     def __post_init__(self):
...         self.y = 2*self.x
...
>>> c = Coor(1) # y 没有初始化
>>> c
Coor(x=1, y=2)
```

总而言之，数据类是一项新功能，它们最终将如何适应常见的工作流还有待观察。这些数据类的灵感来自第三方 attrs 模块，因此请阅读该模块以了解其使用案例是否适用于你的问题。

2.14 泛型函数

泛型函数是根据输入的类型来改变它们的实现方法的函数。例如，你可以在函数的开头使用以下条件语句来完成相同的事情，如下所示：

```
>>> def foo(x):
...     if isinstance(x,int):
...         return 2*x
...     elif isinstance(x,list):
...         return [i*2 for i in x]
...     else:
...         raise NotImplementedError
...
```

用法如下：

```
>>> foo(1)
2
>>> foo([1,2])
[2, 4]
```

在这种情况下，你可以将 foo 视为泛型函数。为了提高这种模式的可靠性，自 Python 3.3 以来，我们可以使用 functools.singledispatch。首先，需要定义上层函数，该函数将根据第一个参数的类型对各个实现进行模板化。

```
>>> from functools import singledispatch
>>> @singledispatch
... def foo(x):
...     print('I am done with type(x): %s'%(str(type(x))))
...
```

相应的输出为：

```
>>> foo(1)
I am done with type(x): <class 'int'>
```

为了使调度正常工作，我们必须使用输入类型作为装饰器参数将新实现 register 到 foo 中。可以将函数命名为 _，因为不需要单独的名称。

```
>>> @foo.register(int)
... def _(x):
...     return 2*x
...
```

现在，让我们再次尝试输出。注意，新的函数 int 版本已经执行。

```
>>> foo(1)
2
```

通过对不同的类型参数再次使用 register，我们可以附加更多基于类型的实现。

```
>>> @foo.register(float)
... def _(x):
...     return 3*x
...
>>> @foo.register(list)
... def _(x):
...     return [3*i for i in x]
...
```

相应的输出为

```
>>> foo(1.3)
3.9000000000000004
>>> foo([1,2,3])
[3, 6, 9]
```

现有的函数还可以使用装饰器的函数形式来附加。

```
>>> def existing_function(x):
...     print('I am the existing_function with %s'%(str(type(x))))
...
>>> foo.register(dict,existing_function)
<function existing_function at 0x7f9398354a60>
```

相应的输出为

```
>>> foo({1:0,2:3})
I am the existing_function with <class 'dict'>
```

你可以看到通过 foo.registry.keys() 实现的 dispatch，如下所示：

```
>>> foo.registry.keys()
dict_keys([<class 'object'>, <class 'int'>, <class 'float'>,
 <class 'list'>, <class 'dict'>])
```

你可以通过访问 dispatch 来选择各个函数，如下所示：

```
>>> foo.dispatch(int)
<function _ at 0x7f939a66b5e0>
```

这些 register 装饰器也可以堆叠并与抽象基类一起使用。

> **编程技巧：使用 slots 减少内存**
>
> 下面的类定义允许任意添加属性。
>
> ```
> >>> class Foo:
> ...     def __init__(self,x):
> ...         self.x=x
> ...
> >>> f = Foo(10)
> >>> f.y = 20
> >>> f.z = ['some stuff', 10,10]
> >>> f.__dict__
> {'x': 10, 'y': 20, 'z': ['some stuff', 10, 10]}
> ```
>
> 这是因为 Foo 中有一个字典，它会产生内存开销，特别是对于许多这样的对象。可以通过添加如下所示的 __slots__ 来消除此开销：
>
> ```
> >>> class Foo:
> ...     __slots__ = ['x']
> ...     def __init__(self,x):
> ...         self.x=x
> ...
> >>> f = Foo(10)
> >>> f.y = 20
> Traceback (most recent call last):
>   File "<stdin>", line 1, in <module>
> AttributeError: 'Foo' object has no attribute 'y'
> ```
>
> 这里触发了 AttributeError，这是因为 __slots__ 阻止了 Python 为每个 Foo 实例创建内部字典。

2.15 设计模式

设计模式在 Python 中并不像在 Java 或 C++ 中那样流行，因为 Python 拥有一个功能广泛且实用的标准库。设计模式代表常见问题的规范解决方案。这个术语来源于建筑学。例如，假设你有一座房子，你的问题是如何在拎着一袋杂货的情况下进入房子。这个问题的解决方案就是门的模式，但这并不指定门的形状或大小，它的颜色，或者它是否有锁等。这些被称为实现细节，主要思想是对于常见问题存在着规范的解决方案。

2.15.1 模板

模板方法是一种行为设计模式，它在基类中定义了算法的骨架，但允许子类覆盖算法的特定步骤而不改变其结构。其动机是将算法分解为一系列步骤，其中每个步骤都有一个抽象的实现。具体来说，这意味着算法的实现细节留给子类，而基类策划算法的各个步骤。

```
>>> from abc import ABC, abstractmethod
>>> # 确保它必须是子类,而不是直接实例
>>> class Algorithm(ABC):
...     # 基类方法
...     def compute(self):
...         self.step1()
...         self.step2()
...     # 子类必须实现这些抽象方法
...     @abstractmethod
...     def step1(self):
...         'step 1 implementation details in subclass'
...         pass
...     @abstractmethod
...     def step2(self):
...         'step 2 implementation details in subclass'
...         pass
...
```

如果你试图直接实例化这个对象，Python 会抛出一个 TypeError。要使用该类，则必须按如下所示对其进行子类化：

```
>>> class ConcreteAlgorithm(Algorithm):
...     def step1(self):
...         print('in step 1')
...     def step2(self):
...         print('in step 2')
...
>>> c = ConcreteAlgorithm()
>>> c.compute()  # compute 在基类定义中
in step 1
in step 2
```

模板的优势在于它清楚地表明由基类协调子类实现的细节。这清楚地分离了关注点，使其能够灵活地在不同情况下部署相同的算法。

2.15.2 单列模式

单列模式是一种创建型设计模式，用于确保一个类只有一个实例。例如，可以有多个打印机，但只有一个打印机后台处理程序。以下代码将单一实例隐藏在

类变量中，并在调用 __init__ 之前使用 __new__ 方法自定义对象创建。

```
>>> class Singleton:
...     # 类变量包含唯一的_instance
...     # __new__ 方法返回指定类的对象,并在 __init__ 被调用之前调用
...     def __new__(cls, *args, **kwds):
...         if not hasattr(cls, '_instance'):
...             cls._instance = super().__new__(cls, *args, **kwds)
...         return cls._instance
...
>>> s = Singleton()
>>> t = Singleton()
>>> t is s # 只有一个实例
True
```

注意，在 Python 中实现此模式的方法有很多，但这是最简单的方法之一。

2.15.3 观察者

观察者模式是一种行为型设计模式，定义了对象之间的通信方式，以便当一个对象（即发布者）改变状态时，所有订阅者都会自动收到通知并更新。traitlets 模块实现了这种设计模式。

```
>>> from traitlets import HasTraits, Unicode, Int
>>> class Item(HasTraits):
...     count = Int()        # 整数的发布者
...     name = Unicode()     # unicode 字符串的发布者
...
>>> def func(change):
...     print('old value of count = ',change.old)
...     print('new value of count = ',change.new)
...
>>> a = Item()
>>> # func 订阅'count'的变化
>>> a.observe(func, names=['count'])
>>> a.count = 1
old value of count =  0
new value of count =  1
>>> a.name = 'abc'   # 不会打印任何内容,因为没有订阅 name
```

设置好以上内容后，我们可以有多个订阅者来订阅发布的属性。

```
>>> def another_func(change):
...     print('another_func is subscribed')
...     print('old value of count = ',change.old)
...     print('new value of count = ',change.new)
...
>>> a.observe(another_func, names=['count'])
>>> a.count = 2
old value of count =  1
new value of count =  2
another_func is subscribed
old value of count =  1
new value of count =  2
```

此外，traitlets 模块对对象属性进行类型检查，如果属性设置了错误的类型，则会引发异常，从而实现 descriptor 模式。traitlets 模块是 Jupyter ipywidgets 生态系统基于 web 的交互式功能的基础。

2.15.4 适配器

适配器模式通过使用类来模拟相关接口，有助于重用现有代码。这允许原本由于接口不兼容而无法相互操作的类进行交互。考虑下面接受列表的类：

```
>>> class EvenFilter:
...     def __init__(self,seq):
...         self._seq = seq
...     def report(self):
...         return [i for i in self._seq if i%2==0]
...
```

这只返回如下所示的偶数项：

```
>>> EvenFilter([1,3,4,5,8]).report()
[4, 8]
```

但现在我们想要使用相同的类，其中输入的 seq 是一个生成器，而不是一个列表。可以使用以下的 GeneratorAdapter 类来实现：

```
>>> class GeneratorAdapter:
...     def __init__(self,gen):
...         self._seq = list(gen)
...     def __iter__(self):
...         return iter(self._seq)
...
```

现在我们可以回过头来使用这个适配器类来配合 EvenFilter，如下所示：

```
>>> g = (i for i in range(10)) # 使用生成器表达式创建生成器
>>> EvenFilter(GeneratorAdapter(g)).report()
[0, 2, 4, 6, 8]
```

适配器模式的主要思想是隔离相关接口并用适配器类模拟它们。

参考文献

1. D.M. Beazley, *Python Essential Reference* (Addison-Wesley, Boston, 2009)
2. A.A. Hagberg, D.A. Schult, P.J. Swart, Exploring network structure, dynamics, and function using NetworkX, in *Proceedings of the 7th Python in Science Conference (SciPy2008)*, Pasadena, CA, August 2008, pp. 11–15

第 3 章

使用模块

模块化保证了代码的重用性和可移植性。通常情况下，广泛使用且经过大量测试的代码要比从头开始编写自己的代码更好。import 语句是将模块加载到当前命名空间中的方法，示例如下：

`import some_module`

为了导入这个模块，Python 会按照 sys.path 目录列表中的条目顺序搜索有效的 Python 模块。PYTHONPATH 环境变量中的项目也会添加到这个搜索路径中。Python 的编译方式会影响导入过程。一般来说，Python 会在文件系统中搜索模块，但某些模块可能会直接编译到 Python 中，这意味着它知道在不搜索文件系统的情况下加载它们的位置。当在共享文件系统上启动数千个 Python 进程时，这可能会导致巨大的性能影响，因为文件系统在搜索过程中可能会导致显著的启动延迟。

3.1 标准库

Python 是一种"自带电池"的语言，意味着许多优秀的模块已经包含在基本语言中。由于它作为 Web 编程语言的历史遗留问题，大多数标准库涉及网络协议和其他对 Web 开发重要的主题。标准库模块在主 Python 网站上有文档可供参考。下面以内置的 math 模块为例：

```
>>> import math          # 导入 math 模块
>>> dir(math)            # 提供模块属性列表
['__doc__', '__file__', '__loader__', '__name__', '__package__',
'__spec__', 'acos', 'acosh', 'asin', 'asinh', 'atan', 'atan2',
'atanh', 'ceil', 'comb', 'copysign', 'cos', 'cosh', 'degrees',
'dist', 'e', 'erf', 'erfc', 'exp', 'expm1', 'fabs', 'factorial',
'floor', 'fmod', 'frexp', 'fsum', 'gamma', 'gcd', 'hypot', 'inf',
'isclose', 'isfinite', 'isinf', 'isnan', 'isqrt', 'ldexp',
'lgamma', 'log', 'log10', 'log1p', 'log2', 'modf', 'nan', 'perm',
'pi', 'pow', 'prod', 'radians', 'remainder', 'sin', 'sinh',
'sqrt', 'tan', 'tanh', 'tau', 'trunc']
>>> help(math.sqrt)
Help on built-in function sqrt in module math:

sqrt(x, /)
    Return the square root of x.

>>> radius = 14.2
>>> area = math.pi*(radius**2)
>>> area                       # 使用模块变量
633.4707426698459
>>> a = 14.5; b = 12.7
>>> c = math.sqrt(a**2+b**2)   # 使用模块函数
>>> c
19.275372888740698
```

一旦一个模块被导入，它就会包含在 sys.modules 字典中。导入的第一步是检查这个字典是否包含所需的模块，如果已经包含，则不会重新导入它。这意味着在解释器中多次执行 import somemodule 并不会重新加载模块。这是因为导入解析协议的第一步是查看 sys.modules 字典中是否已经存在所需的模块，如果已经包含，则不会再次导入它。

有时你只需要从给定模块中使用特定函数。这种情况可以使用"from <module> import <name>"语法处理。

```
>>> # 会覆盖命名空间中现有的定义
>>> from math import sqrt
>>> sqrt(a) # 简便
3.8078865529319543
```

注意，你必须使用 importlib.reload 函数将模块重新导入工作区。重要的是，importlib.reload 不适用于上面使用的 from 语法。因此，如果你正在开发代码并不断重新加载它，最好保留顶级模块名称，这样就可以使用 importlib.reload 不断重新加载它。

在导入过程中有许多地方可以进行劫持操作。这使得创建虚拟环境和定制执行变得可能，但在这些情况下，最好使用已经成熟的解决方案（参见下文中的 conda）。

3.2 编写和使用自己的模块

除了使用极其优秀的标准库之外，你可能还想与同事分享你的代码。最简单的方法是将代码放在一个文件中并发送给他们。当交互式开发要以这种方式分发代码时，你必须了解你的模块在活动解释器中是如何更新（或不更新）的。

例如，将以下代码放在一个名为 mystuff.py 的单独文件中：

```python
def sqrt(x):
    return x*x
```

然后返回交互式会话。

```
>>> from mystuff import sqrt
>>> sqrt(3)
9
>>> import mystuff
>>> dir(mystuff)
['__builtins__', '__cached__', '__doc__',
 '__file__','__loader__','__name__','__package__',
 '__spec__', 'sqrt']
>>> mystuff.sqrt(3)
9
```

现在，将以下函数添加到 mystuff.py 文件中：

```python
def poly_func(x):
    return 1+x+x*x
```

并更改文件中以前的函数。

```python
def sqrt(x):
    return x/2
```

然后返回交互式会话。

```
mystuff.sqrt(3)
```

能得到想要的结果吗？

```
mystuff.poly_func(3)
```

注意，你必须使用 importlib.reload 才能将文件中的新更改导入解释器。一个名为 __pycache__ 的目录会自动出现在与 mystuff.py 相同的目录中。这是 Python 存储模块编译后的字节码的地方，这样 Python 就不必每次导入 mystuff.py 时都重新从头编译它。每当你对 mystuff.py 进行更改时，此目录将自动刷新。永远不要将 __pycache__ 目录包含在你的 Git 存储库中，因为当其他人克隆你的存储库时，如果文件系统的时间戳出错，可能会导致 __pycache__ 与源代码不同步。这是一个令人痛苦的 bug，其他人可能对 mystuff.py 文件进行更改，但当导入 mystuff 模块时，这些更改将不会生效，因为 Python 仍在使用 __pycache__ 中的版本。如果你正在使用 Python 2.x，那么编译后的 Python 字节码会以 .pyc 文件的形式存储在同一目录中，而不是在 __pycache__ 中。出于同样的原因，这些文件也永远不应包含在你的 Git 存储库中。

> **编程技巧：IPython 自动重新加载**
>
> IPython 通过 %autoreload 提供了一些自动重新加载功能，但它附带了很多需要注意的事项。因此，最好的方法还是去明确地使用 importlib.reload。

3.2.1　将目录用作模块

除了将所有的 Python 代码放入单个文件之外，你可以使用目录将代码组织成单独的文件。诀窍是在你想要从中导入的目录的顶层放置一个名为 __init__.py 的文件。该文件可以为空。例如：

```
package/
    __init__.py
    moduleA.py
```

如果 package 在你的路径中，你可以 import package。如果 __init__.py 是空的，那么它什么也不做。要获取 moduleA.py 中的任何代码，你必须显式地导入它，import package.moduleA。例如，package.moduleA.foo( ) 会运行 moduleA.py 文件中的 foo 函数。如果你想在导入 package 时使 foo 可用，那么必须在 __init__.py 文件中放置 from.moduleA import foo。在 Python 3 中需要使用相对导入符号。然后，你可以执行 import package，然后运行函数 package.foo( )。你还可以执行 from package import foo 直接获取 foo。在开发自己的模块时，可以使用相对导入来对导入的包进行细粒度控制。

3.3　动态导入

如果你不知道需要提前导入的模块的名称，那么可以使用 __import__ 函数从指定的模块列表中加载模块。

```
>>> sys = __import__('sys')   # 从字符串参数导入模块
>>> sys.version
'3.8.3 (default, May 19 2020, 18:47:26) \n[GCC 7.3.0]'
```

> **编程技巧：使用 __main__**
>
> 命名空间用于区分导入和运行 Python 脚本。
>
> ```
> if __name__ == '__main__':
>     # 这些语句在导入时不会被执行
>     # 在这里运行语句
> ```
>
> 还有 __file__ 属性，它表示被导入文件的文件名。除了 __init__.py 文件，如果你希望通过命令行使用 -m 开关调用模块，可以在模块目录的顶层添加一个 __main__.py 文件。这样做意味着 __init__.py 文件中的代码也会运行。

3.4 从 Web 中获取模块

Python 的打包在过去几年中确实有了很大的改进。这曾经是一个痛点，但现在部署和维护依赖于多个平台的链接库的 Python 代码变得更加容易了。

```
% pip install name_of_module
```

Python 支持使用 pip 找出并安装所有相关模块，还可以控制包的安装方式和位置。你不需要根访问权限就可以使用它，可参阅非根访问的 --user。现代 Python 打包依赖于 wheel 文件，其中包含模块所依赖的编译库的片段。通常情况下，这种方式能够正常工作，但如果你在 Windows 平台上遇到问题，可以参考加州大学欧文分校 Christoph Gohlke 的实验室⊖提供的大量适用于 Windows 的 wheel 文件。

3.5 Conda 包管理

conda 可以缓解许多软件包管理方面的困扰，并且不需要管理员 / 根权限来有效地使用它。anaconda 工具集是由 Anaconda 公司支持的科学软件包的列表，它几乎包含了你想要的所有科学软件包。除此之外，社区还支持 conda-forge。你可以使用以下方式将 conda-forge 添加到常规存储库列表中：

⊖ 参见 https://www.l.uci.edu/~gohlke/pythonlibs。

```
Terminal> conda config --add channels conda-forge
```
然后，你可以按以下方式安装新模块：
```
Terminal> conda install name_of_module
```
此外，conda 还支持创建自包含的子环境，这是一种安全地尝试代码甚至不同 Python 版本的好方法。这对于在云计算环境中自动配置虚拟机非常重要。例如，以下命令将创建一个名为 my_test_env 的环境，其中包含 Python 版本为 3.7。
```
Terminal> conda create -n my_test_env python=3.7
```
pip 和 conda 的区别在于，pip 将根据所需包的要求来确保缺失模块的安装。conda 包管理器也会执行相同的操作，但还会确定所需包及其依赖项的版本与现有安装之间是否存在冲突，并提前发出警告[一]。这避免了为满足新安装而覆盖预先存在的包的依赖关系的问题。最佳的方法是在处理具有许多链接库的科学代码时首选 conda，然后在处理纯 Python 代码时使用 pip。有时需要同时使用两者，因为某些所需的 Python 模块可能尚未得到 conda 的支持。这可能会出现问题，因为 conda 无法将通过 pip 安装的新包纳入其内部管理。有关更多信息，可参考 conda 文档，并注意 conda 仍在不断开发中。

创建虚拟环境的另一种方法是使用 venv（或 virtualenv），它是 Python 自带的。同样，这对于使用 pip 安装的包和纯 Python 代码是一个不错的选择，但对于科学程序来说，conda 是更好的选择。然而，使用 venv 或 virtualenv 创建的虚拟环境特别适用于隔离命令行程序，这些程序可能具有你不希望携带或干扰其他安装的奇怪依赖项。

编程技巧：Mamba 包管理器

Mamba 包管理器比 conda 快数倍，因为它有更高效的可满足性求解器，是 conda 的优秀替代品。

[一] 为了解决冲突，conda 采用了一个可满足性求解器（SAT），这是一个经典的组合问题。

参考文献

1. D.M. Beazley, *Python Essential Reference* (Addison-Wesley, Boston, 2009)
2. D. Beazley, B.K. Jones, *Python Cookbook: Recipes for Mastering Python 3* (O'Reilly Media, Newton, 2013)
3. N. Ceder, *The Quick Python Book.* (Manning Publications, Shelter Island, 2018)
4. D. Hellmann, *The Python 3 Standard Library by Example* Developer's Library (Pearson Education, London, 2017)
5. C. Hill, *Learning Scientific Programming With Python* (Cambridge University Press, Cambridge, 2020)
6. D. Kuhlman, *A Python Book: Beginning Python, Advanced Python, and Python Exercises* (Platypus Global Media, Washington, 2011)
7. H.P. Langtangen, *A Primer on Scientific Programming With Python.* Texts in Computational Science and Engineering (Springer, Berlin, Heidelberg, 2016)
8. M. Lutz, *Learning Python: Powerful Object-Oriented Programming.* Safari Books Online (O'Reilly Media, Newton, 2013)
9. M. Pilgrim, *Dive Into Python 3.* Books for Professionals by Professionals (Apress, New York, 2010)
10. K. Reitz, T. Schlusser, *The Hitchhiker's Guide to Python: Best Practices for Development* (O'Reilly Media, Newton, 2016)
11. C. Rossant, *Learning IPython for Interactive Computing and Data Visualization* (Packt Publishing, Birmingham, 2015)
12. Z.A. Shaw, *Learn Python the Hard Way: Release 2.0.* Lulu.com (2012)
13. M. Summerfield, Python in Practice: Create Better Programs Using Concurrency, Libraries, and Patterns (Pearson Education, London, 2013)
14. J. Unpingco, *Python for Signal Processing: Featuring IPython Notebooks* (Springer International Publishing, Cham, 2016)
15. J. Unpingco, *Python for Probability, Statistics, and Machine Learning*, 2nd edn. (Springer International Publishing, Cham, 2019)

第 4 章

Numpy

Numpy 提供了一种统一的方式来管理 Python 中的数值数组。它整合了许多先前在 Python 中处理数值数组的方法中的最佳思想，是许多其他科学计算 Python 模块的基石。想要理解和有效地使用科学计算和 Python，对 Numpy 的扎实掌握是必不可少的。

```
>>> import numpy as np # 命名规则
```

4.1 Dtypes

虽然 Python 是动态类型的，但 Numpy 允许使用 dtypes 对数字类型进行精确规范。

```
>>> a = np.array([0],np.int16) # 16bit 整数
>>> a.itemsize # 元素字节长度(8位字节)
2
>>> a.nbytes
2
>>> a = np.array([0],np.int64) # 64bit 整数
>>> a.itemsize
8
```

数组遵循相同的模式，

```
>>> a = np.array([0,1,23,4],np.int64) # 64bit 整数
>>> a.shape
```

```
(4,)
>>> a.nbytes
32
```

注意，Numpy 数组创建后就无法再添加额外元素。

```
>>> a = np.array([1,2])
>>> a[2] = 32
Traceback (most recent call last):
  File "<stdin>", line 1, in <module>
IndexError: index 2 is out of bounds for axis 0 with size 2
```

这是因为内存块已经被划分，并且 Numpy 不会在没有明确指令的情况下分配新内存并复制数据。此外，一旦你使用特定的 dtype 创建数组，对该数组进行赋值将会转换为该类型。例如：

```
>>> x = np.array(range(5), dtype=int)
>>> x[0] = 1.33 # 浮点赋值与 dtype=int 不匹配
>>> x
array([1, 1, 2, 3, 4])
>>> x[0] = 'this is a string'
Traceback (most recent call last):
  File "<stdin>", line 1, in <module>
ValueError: invalid literal for int() with base 10: 'this is a
↪   string'
```

4.2 多维数组

多维数组遵循相同的模式。

```
>>> a = np.array([[1,3],[4,5]]) # 省略数据类型将选择默认值
>>> a
array([[1, 3],
       [4, 5]])
>>> a.dtype
dtype('int64')
>>> a.shape
(2, 2)
>>> a.nbytes
32
>>> a.flatten()
array([1, 3, 4, 5])
```

维度的最大限制取决于 Numpy 构建期间的配置（通常为 32）。Numpy 提供了许多自动构建数组的方法。

```
>>> a = np.arange(10) # 类似于 range()
>>> a
array([0, 1, 2, 3, 4, 5, 6, 7, 8, 9])
>>> a = np.ones((2,2))
>>> a
array([[1., 1.],
       [1., 1.]])
```

```
>>> a = np.linspace(0,1,5)
>>> a
array([0.  , 0.25, 0.5 , 0.75, 1.  ])
>>> X,Y = np.meshgrid([1,2,3],[5,6])
>>> X
array([[1, 2, 3],
       [1, 2, 3]])
>>> Y
array([[5, 5, 5],
       [6, 6, 6]])
>>> a = np.zeros((2,2))
>>> a
array([[0., 0.],
       [0., 0.]])
```

你还可以使用函数创建 Numpy 数组。

```
>>> np.fromfunction(lambda i,j: abs(i-j)<=1, (4,4))
array([[ True,  True, False, False],
       [ True,  True,  True, False],
       [False,  True,  True,  True],
       [False, False,  True,  True]])
```

Numpy 数组也可以有字段名。

```
>>> a = np.zeros((2,2), dtype=[('x','f4')])
>>> a['x']
array([[0., 0.],
       [0., 0.]], dtype=float32)
>>> x = np.array([(1,2)], dtype=[('value','f4'),
...                              ('amount','c8')])
>>> x['value']
array([1.], dtype=float32)
>>> x['amount']
array([2.+0.j], dtype=complex64)
>>> x = np.array([(1,9),(2,10),(3,11),(4,14)],
...              dtype=[('value','f4'),
...                     ('amount','c8')])
>>> x['value']
array([1., 2., 3., 4.], dtype=float32)
>>> x['amount']
array([ 9.+0.j, 10.+0.j, 11.+0.j, 14.+0.j], dtype=complex64)
```

Numpy 数组也可以使用 recarray 通过其属性进行访问。

```
>>> y = x.view(np.recarray)
>>> y.amount # 作为属性访问
array([ 9.+0.j, 10.+0.j, 11.+0.j, 14.+0.j], dtype=complex64)
>>> y.value # 作为属性访问
array([1., 2., 3., 4.], dtype=float32)
```

4.3 重塑和堆叠 Numpy 数组

数组可以水平和垂直堆叠。

```
>>> x = np.arange(5)
>>> y = np.array([9,10,11,12,13])
>>> np.hstack([x,y]) # 水平堆叠
array([ 0,  1,  2,  3,  4,  9, 10, 11, 12, 13])
>>> np.vstack([x,y]) # 垂直堆叠
array([[ 0,  1,  2,  3,  4],
       [ 9, 10, 11, 12, 13]])
```

如果要在第三个维度 depth 中进行叠加，可以使用 dstack 方法。Numpy np.concatenate 处理通常的任意维度情况。在某些代码（例如，scikit-learn）中，可能会发现更简洁的 np.c_ 和 np.r_ 用于按列和按行堆叠数组。

```
>>> np.c_[x,x] # 按列
array([[0, 0],
       [1, 1],
       [2, 2],
       [3, 3],
       [4, 4]])
>>> np.r_[x,x] # 按行
array([0, 1, 2, 3, 4, 0, 1, 2, 3, 4])
```

4.4 复制 Numpy 数组

Numpy 有一个复制元素的 repeat 函数，还有一个更通用的 tile 函数可指定块矩阵的形状。

```
>>> x=np.arange(4)
>>> np.repeat(x,2)
array([0, 0, 1, 1, 2, 2, 3, 3])
>>> np.tile(x,(2,1))
array([[0, 1, 2, 3],
       [0, 1, 2, 3]])
>>> np.tile(x,(2,2))
array([[0, 1, 2, 3, 0, 1, 2, 3],
       [0, 1, 2, 3, 0, 1, 2, 3]])
```

你也可以将字符串等非数字作为数组中的项。

```
>>> np.array(['a','b','cow','deep'])
array(['a', 'b', 'cow', 'deep'], dtype='<U4')
```

注意，'U4' 是指长度为 4 的字符串，它是序列中最长的字符串。

重塑 Numpy 数组

Numpy 数组可以在创建后重塑。

```
>>> a = np.arange(10).reshape(2,5)
>>> a
array([[0, 1, 2, 3, 4],
       [5, 6, 7, 8, 9]])
```

对于真正懒惰的人，你可以用负的维度替换上面的一个维度（即 reshape

(−1, 5)),Numpy 将计算出符合条件的另一个维度。数组 transpose 方法的操作与 .T 属性相同。

```
>>> a.transpose()
array([[0, 5],
       [1, 6],
       [2, 7],
       [3, 8],
       [4, 9]])
>>> a.T
array([[0, 5],
       [1, 6],
       [2, 7],
       [3, 8],
       [4, 9]])
```

共轭转置（即 Hermitian 转置）是 .H 属性。

4.5 切片、逻辑数组操作

Numpy 数组遵循与 Python 列表和字符串相同的零索引切片逻辑。

```
>>> x = np.arange(50).reshape(5,10)
>>> x
array([[ 0,  1,  2,  3,  4,  5,  6,  7,  8,  9],
       [10, 11, 12, 13, 14, 15, 16, 17, 18, 19],
       [20, 21, 22, 23, 24, 25, 26, 27, 28, 29],
       [30, 31, 32, 33, 34, 35, 36, 37, 38, 39],
       [40, 41, 42, 43, 44, 45, 46, 47, 48, 49]])
```

冒号意味着沿着指示的维度取走所有元素。

```
>>> x[:,0]
array([ 0, 10, 20, 30, 40])
>>> x[0,:]
array([0, 1, 2, 3, 4, 5, 6, 7, 8, 9])
>>> x = np.arange(50).reshape(5,10) # 重塑数组
>>> x
array([[ 0,  1,  2,  3,  4,  5,  6,  7,  8,  9],
       [10, 11, 12, 13, 14, 15, 16, 17, 18, 19],
       [20, 21, 22, 23, 24, 25, 26, 27, 28, 29],
       [30, 31, 32, 33, 34, 35, 36, 37, 38, 39],
       [40, 41, 42, 43, 44, 45, 46, 47, 48, 49]])
>>> x[:,0] # 所有行,第 0 列
array([ 0, 10, 20, 30, 40])
>>> x[0,:] # 所有列,第 0 行
array([0, 1, 2, 3, 4, 5, 6, 7, 8, 9])
>>> x[1:3,4:6]
array([[14, 15],
       [24, 25]])
>>> x = np.arange(2*3*4).reshape(2,3,4) # 重塑数组
>>> x
array([[[ 0,  1,  2,  3],
        [ 4,  5,  6,  7],
        [ 8,  9, 10, 11]],
```

```
       [[12, 13, 14, 15],
        [16, 17, 18, 19],
        [20, 21, 22, 23]]])
>>> x[:,1,[2,1]]  # 索引每个维度
array([[ 6,  5],
       [18, 17]])
```

Numpy 的 where 函数可以根据特定的逻辑条件查找数组元素。注意，np.where 返回 Numpy 索引的元组。

```
>>> np.where(x % 2 == 0)
(array([0, 0, 0, 0, 0, 0, 1, 1, 1, 1, 1, 1]),
array([0, 0, 1, 1, 2, 2, 0, 0, 1, 1, 2, 2]),
array([0, 2, 0, 2, 0, 2, 0, 2, 0, 2, 0, 2]))
>>> x[np.where(x % 2 == 0)]
array([ 0,  2,  4,  6,  8, 10, 12, 14, 16, 18, 20, 22])
>>> x[np.where(np.logical_and(x % 2 == 0,x < 9))]  # 还包括 logical_or 等
array([0, 2, 4, 6, 8])
```

此外，Numpy 数组可以通过逻辑 Numpy 数组进行索引，其中只选择相应的 True 条目。

```
>>> a = np.arange(9).reshape((3,3))
>>> a
array([[0, 1, 2],
       [3, 4, 5],
       [6, 7, 8]])
>>> b = np.fromfunction(lambda i,j: abs(i-j) <= 1, (3,3))
>>> b
array([[ True,  True, False],
       [ True,  True,  True],
       [False,  True,  True]])
>>> a[b]
array([0, 1, 3, 4, 5, 7, 8])
>>> b = (a>4)
>>> b
array([[False, False, False],
       [False, False,  True],
       [ True,  True,  True]])
>>> a[b]
array([5, 6, 7, 8])
```

4.6　Numpy 数组和内存

Numpy 使用引用传递（pass-by-reference）语义，因此切片操作返回的是数组的视图，而不是隐式复制数据，这与 Python 的语义保持一致。这对已经占用大量内存的大型数组特别有用。在 Numpy 的术语中，切片操作创建视图（不复制数据），而高级索引操作会创建副本。让我们从高级索引开始。

如果索引对象（即方括号之间的项目）是非元组序列对象、另一个 Numpy 数组（整数或布尔类型），或者至少包含一个序列对象或 Numpy 数组的元组，则索引会创建副本。要扩展和复制 Numpy 中的现有数组，需要执行类似以下操作：

```
>>> x = np.ones((3,3))
>>> x
array([[1., 1., 1.],
       [1., 1., 1.],
       [1., 1., 1.]])
>>> x[:,[0,1,2,2]] # 注意最后一个维度的重复
array([[1., 1., 1., 1.],
       [1., 1., 1., 1.],
       [1., 1., 1., 1.]])
>>> y=x[:,[0,1,2,2]] # 同上,但请将其指定给 y
```

由于高级索引，x 的相关部分被复制，因此变量 y 拥有自己的内存空间。为了证明这一点，我们给 x 分配了一个新元素，并看到 y 没有更新。

```
>>> x[0,0]=999 # 更改 x 中的元素
>>> x # 改变了
array([[999.,    1.,    1.],
       [  1.,    1.,    1.],
       [  1.,    1.,    1.]])
>>> y # 无变化。
array([[1., 1., 1., 1.],
       [1., 1., 1., 1.],
       [1., 1., 1., 1.]])
```

然而，如果我们重新开始并通过如下所示的切片来构造 y，那么对 x 所做的更改确实会影响 y，因为视图只是在同一内存中的一个窗口。

```
>>> x = np.ones((3,3))
>>> y = x[:2,:2] # 左上部分视图
>>> x[0,0] = 999 # 更改 x 中的元素
>>> x   # 看到变化了吗?
array([[999.,    1.,    1.],
       [  1.,    1.,    1.],
       [  1.,    1.,    1.]])
>>> y
array([[999.,    1.],
       [  1.,    1.]])
```

注意，如果要显式强制复制而不使用任何索引技巧，则可以执行 y=x.copy( )。下面的代码是高级索引与切片的另一个示例：

```
>>> x = np.arange(5) # 创建数组
>>> x
array([0, 1, 2, 3, 4])
>>> y=x[[0,1,2]] # 按整数列表索引以强制复制
>>> y
array([0, 1, 2])
>>> z=x[:3]          # 切片创建视图
```

```
>>> z                # 注意 y 和 z 具有相同的条目
array([0, 1, 2])
>>> x[0]=999         # 更改 x 的元素
>>> x
array([999,   1,   2,   3,   4])
>>> y                # 注意 y 不受影响
array([0, 1, 2])
>>> z                # 但是 z 受影响(它是一个视图)
array([999,   1,   2])
```

在本例中，y 是副本，而不是视图，因为它是使用高级索引创建的，而 z 是使用切片创建的。因此，即使 y 和 z 具有相同的条目，也只有 z 受 x 的更改影响。

重叠 Numpy 数组

使用视图来操作内存对于需要重叠内存片段的信号和图像处理算法特别有用。以下是一个示例，展示了如何使用高级 Numpy 创建不实际消耗额外内存的重叠块：

```
>>> from numpy.lib.stride_tricks import as_strided
>>> x = np.arange(16).astype(np.int32)
>>> y=as_strided(x,(7,4),(8,4)) # 重叠
>>> y
array([[ 0,  1,  2,  3],
       [ 2,  3,  4,  5],
       [ 4,  5,  6,  7],
       [ 6,  7,  8,  9],
       [ 8,  9, 10, 11],
       [10, 11, 12, 13],
       [12, 13, 14, 15]], dtype=int32)
```

以上代码创建了一系列整数，然后通过重叠条目创建了一个 7×4 的 Numpy 数组。as_strided 函数中的最后一个参数是步幅（strides），分别表示在行和列维度上移动的字节步长。因此，结果数组在列维度上每次移动四个字节，在行维度上每次移动八个字节。由于 Numpy 数组中的整数元素占据四个字节，这相当于在列维度上移动一个元素，在行维度上移动两个元素。Numpy 数组的第二行从第一个条目（即 2）的 8 个字节（两个元素）开始，然后在列维度上以四个字节（一个元素）的步长继续（即 2，3，4，5）。重要的是，内存在生成的 7×4 Numpy 数组中被重复使用。下面的代码通过重新分配原始 x 数组中的元素来证明这一点。更改会在 y 数组中显示出来，因为它们指向相同的分配内存。

```
>>> x[::2] = 99 # 每隔一个值赋值
>>> x
array([99,  1, 99,  3, 99,  5, 99,  7, 99,  9, 99, 11, 99, 13, 99,
       15],
      dtype=int32)
>>> y # 因为 y 是一个视图,所以会出现更改
array([[99,  1, 99,  3],
```

```
 [99,  3, 99,  5],
 [99,  5, 99,  7],
 [99,  7, 99,  9],
 [99,  9, 99, 11],
 [99, 11, 99, 13],
 [99, 13, 99, 15]], dtype=int32)
```

注意，as_strided 函数不会检查你是否保持在内存块边界内。因此，如果目标矩阵的大小未被可用数据填满，剩余的元素将来自该内存位置的任何字节。换句话说，没有默认用零或其他策略填充以保护内存块边界。一种防御方法是显式控制维度。

```
>>> n = 8 # 元素数量
>>> x = np.arange(n) # 创建数组
>>> k = 5 # 行数设定
>>> y = as_strided(x,(k,n-k+1),(x.itemsize,)*2)
>>> y
array([[0, 1, 2, 3],
       [1, 2, 3, 4],
       [2, 3, 4, 5],
       [3, 4, 5, 6],
       [4, 5, 6, 7]])
```

4.7　Numpy 内存数据结构

让我们检查一下 Numpy 源代码中的数据结构 typedef。

```
typedef struct PyArrayObject {
    PyObject_HEAD

        /* Block of memory */
        char *data;

    /* Data type descriptor */
    PyArray_Descr *descr;

    /* Indexing scheme */
    int nd;
    npy_intp *dimensions;
    npy_intp *strides;

    /* Other stuff */
    PyObject *base;
    int flags;
    PyObject *weakreflist;
} PyArrayObject;0
```

让我们创建一个 16 位整数的 Numpy 数组，并对其进行探索。

```
>>> x = np.array([1], dtype=np.int16)
```

我们可以使用 x.data 属性查看原始数据。

```
>>> bytes(x.data)
b'\x01\x00'
```

需注意字节的方向。现在，将数据类型更改为无符号两字节大端整数。

```
>>> x = np.array([1], dtype='>u2')
>>> bytes(x.data)
b'\x00\x01'
```

再次注意字节的方向。这就是小/大端对内存中数据的意义。我们可以使用 frombuffer 直接从字节创建 Numpy 数组，如下所示：

```
>>> np.frombuffer(b'\x00\x01',dtype=np.int8)
array([0, 1], dtype=int8)
>>> np.frombuffer(b'\x00\x01',dtype=np.int16)
array([256], dtype=int16)
>>> np.frombuffer(b'\x00\x01',dtype='>u2')
array([1], dtype=uint16)
```

一旦创建了 ndarray，你可以重新将其转换为不同的 dtype 或更改视图。粗略地说，类型转换会复制数据。例如：

```
>>> x = np.frombuffer(b'\x00\x01',dtype=np.int8)
>>> x
array([0, 1], dtype=int8)
>>> y = x.astype(np.int16)
>>> y
array([0, 1], dtype=int16)
>>> y.flags['OWNDATA']  # y是副本
True
```

或者，我们可以使用 view 重新解释数据。

```
>>> y = x.view(np.int16)
>>> y
array([256], dtype=int16)
>>> y.flags['OWNDATA']
False
```

注意，这里 y 不是新内存，它只是引用现有内存，并使用不同的 dtype 重新解释它。

Numpy 内存步长

上述 typedef 中的步长与 Numpy 如何在数组之间移动有关。步长是到达下一个连续数组元素所需的字节数。每个维度有一个步长。考虑以下 Numpy 数组：

```
>>> x = np.array([[1, 2, 3], [4, 5, 6], [7, 8, 9]], dtype=np.int8)
>>> bytes(x.data)
b'\x01\x02\x03\x04\x05\x06\x07\x08\t'
>>> x.strides
(3, 1)
```

如果我们想要索引 x[1，2]，则必须使用以下偏移：

```
>>> offset = 3*1+1*2
>>> x.flat[offset]
6
```

Numpy 支持 C 顺序（即按列）和 Fortran 顺序（即按行）。例如：

```
>>> x = np.array([[1, 2, 3], [7, 8, 9]], dtype=np.int8,order='C')
>>> x.strides
(3, 1)
>>> x = np.array([[1, 2, 3], [7, 8, 9]], dtype=np.int8,order='F')
>>> x.strides
(1, 2)
```

注意这两种排序的步长之间的差异。对于 C 顺序，行之间移动需要 3 个字节，列之间移动需要 1 个字节，而对于 Fortran 顺序，行之间移动需要 1 个字节，列之间移动需要 2 个字节。这种模式适用于更高的维度。

```
>>> x = np.arange(125).reshape((5,5,5)).astype(np.int8)
>>> x.strides
(25, 5, 1)
>>> x[1,2,3]
38
```

要使用字节偏移量获取 [1, 2, 3] 元素，可以执行以下操作：

```
>>> offset = (25*1 + 5*2 +1*3)
>>> x.flat[offset]
38
```

通过切片创建视图只会改变步长。

```
>>> x = np.arange(3,dtype=np.int32)
>>> x.strides
(4,)
>>> y = x[::-1]
>>> y.strides
(-4,)
```

置换也只是交换步长。

```
>>> x = np.array([[1, 2, 3], [7, 8, 9]], dtype=np.int8,order='F')
>>> x.strides
(1, 2)
>>> y = x.T
>>> y.strides # 负数
(2, 1)
```

一般来说，重塑不仅改变步长，有时还可能复制数据。由于 CPU 缓存，内存布局（即步长）可能会影响性能。CPU 以块的形式从主存中提取数据，这样，如果可以在单个块中连续操作多个项目，则可以减少从主存传输数据的次数，从而加快计算速度。

4.8 数组元素操作

在 Numpy 中，通常的成对算术运算是按元素级的。

```
>>> x*3
array([[ 3,  6,  9],
       [21, 24, 27]], dtype=int8)
>>> y = x/x.max()
>>> y
array([[0.11111111, 0.22222222, 0.33333333],
       [0.77777778, 0.88888889, 1.        ]])
>>> np.sin(y) * np.exp(-y)
array([[0.09922214, 0.17648072, 0.23444524],
       [0.32237812, 0.31917604, 0.30955988]])
```

> **编程技巧：谨慎使用 Numpy 的原位操作**
>
> 对于 Numpy 数组而言，诸如 x-=x.T 这样的原位操作很容易出错，因此通常应避免使用，否则可能导致难以发现的错误。

4.9 通用函数

现在我们知道了如何创建和操作 Numpy 数组，接下来考虑如何使用其他 Numpy 功能进行计算。通用函数（ufuncs）是经过优化的 Numpy 函数，用于在 C 级（即 Python 解释器之外）计算 Numpy 数组。接下来计算三角正弦。

```
>>> a = np.linspace(0,1,20)
>>> np.sin(a)
array([0.        , 0.05260728, 0.10506887, 0.15723948, 0.20897462,
       0.26013102, 0.310567  , 0.36014289, 0.40872137, 0.45616793,
       0.50235115, 0.54714315, 0.59041986, 0.63206143, 0.67195255,
       0.70998273, 0.74604665, 0.78004444, 0.81188195, 0.84147098])
```

注意，Python 有一个内置的数学模块，它有自己的正弦函数。

```
>>> from math import sin
>>> [sin(i) for i in a]
[0.0, 0.05260728333807213, 0.10506887376594912,
0.15723948186175024, 0.20897462406278547, 0.2601310228046501,
0.3105670033203749, 0.3601428860007191, 0.40872137322898616,
0.45616792961904572, 0.5023511546035125, 0.547143146340223,
0.5904198559291864, 0.6320614309590333, 0.6719525474315213,
0.7099827291448582, 0.7460466536513234, 0.7800444439418607,
0.8118819450498316, 0.8414709848078965]
```

输出是一个列表，而不是一个 Numpy 数组，为了处理 a 的所有元素，需要使用列表推导式来计算正弦函数。这是因为 Python 的 math 函数只能逐个处理数组的每个成员。Numpy 的正弦函数不需要这些额外的语义，因为计算是在 Python 解释器之外的 Numpy C 代码中运行的。这就是 Numpy 相对于纯 Python 代码加速 200~300 倍的原因。

```
>>> np.array([sin(i) for i in a])
array([0.        , 0.05260728, 0.10506887, 0.15723948, 0.20897462,
       0.26013102, 0.310567  , 0.36014289, 0.40872137, 0.45616793,
       0.50235115, 0.54714315, 0.59041986, 0.63206143, 0.67195255,
       0.70998273, 0.74604665, 0.78004444, 0.81188195, 0.84147098])
```

上面的操作完全违背了使用 Numpy 的目的。应该尽可能使用 Numpy 的通用函数。

```
>>> np.sin(a)
array([0.        , 0.05260728, 0.10506887, 0.15723948, 0.20897462,
       0.26013102, 0.310567  , 0.36014289, 0.40872137, 0.45616793,
       0.50235115, 0.54714315, 0.59041986, 0.63206143, 0.67195255,
       0.70998273, 0.74604665, 0.78004444, 0.81188195, 0.84147098])
```

4.10　Numpy 数据输入 / 输出

Numpy 可以轻松地将数据移入和移出文件。

```
>>> x = np.loadtxt('sample1.txt')
>>> x
array([[ 0.,  0.],
       [ 1.,  1.],
       [ 2.,  4.],
       [ 3.,  9.],
       [ 4., 16.],
       [ 5., 25.],
       [ 6., 36.],
       [ 7., 49.],
       [ 8., 64.],
       [ 9., 81.]])
>>> #每列类型不同
>>> x = np.loadtxt('sample1.txt',dtype='f4,i4')
>>> x
array([(0.,  0), (1.,  1), (2.,  4), (3.,  9), (4., 16), (5.,
       25),
       (6., 36), (7., 49), (8., 64), (9., 81)],
      dtype=[('f0', '<f4'), ('f1', '<i4')])
```

Numpy 数组可以用相应的 np.savetxt 函数保存。

4.11　线性代数

Numpy 可直接访问经过验证的 LAPACK/BLAS 线性代数代码。在 Numpy 中，线性代数函数的主要入口是通过 linalg 子模块。

```
>>> np.linalg.eig(np.eye(3))  # 运行底层 LAPACK/BLAS
(array([1., 1., 1.]), array([[1., 0., 0.],
       [0., 1., 0.],
       [0., 0., 1.]]))
>>> np.eye(3)*np.arange(3)  # 这是否如预期的那样有效？
array([[0., 0., 0.],
       [0., 1., 0.],
       [0., 0., 2.]])
```

要获得矩阵的行 - 列乘积，可以使用矩阵对象。

```
>>> np.eye(3)*np.matrix(np.arange(3)).T  # 行 - 列乘法
matrix([[0.],
        [1.],
        [2.]])
```

也可以使用 Numpy 的 dot。

```
>>> a = np.eye(3)
>>> b = np.arange(3).T
>>> a.dot(b)
array([0., 1., 2.])
>>> b.dot(b)
5
```

dot 的优点是可以在任意尺寸下工作。这对于类似张量的收缩很方便（有关更多信息，请参见 Numpy 的 tensordot）。自 Python 3.6 以后，还有 @ 表示 Numpy 矩阵乘法。

```
>>> a = np.eye(3)
>>> b = np.arange(3).T
>>> a @ b
array([0., 1., 2.])
```

4.12 广播

广播的功能非常强大，但理解它需要时间。引用 Travis Oliphant 的《A Guide to NumPy》[1] 中的描述：

1）所有维度小于最大维度输入数组的输入数组，在它们的形状前面都会加上 1。

2）输出形状中每个维度的大小是该维度上所有输入形状中的最大值。

3）如果输入在特定维度的形状与输出形状匹配，或者在该维度上的值恰好为 1，则可以在计算中使用该输入。

4）如果输入在其形状中某个维度的大小为 1，则在该维度上所有计算都将使用该维度的第一个数据条目。换句话说，ufunc 的步进机制在需要时将不会沿着该维度步进（该维度的步幅将为 0）。

这些规则的更简单的理解方式如下：

1）如果数组的形状长度不同，则用 1 在左侧填充较小的形状。

2）如果任何对应的维度不匹配，则沿着 1 维度进行复制。

3）如果任何对应的维度不包含 1，则引发错误。

示例如下：

```
>>> x = np.arange(3)
>>> y = np.arange(5)
```

假设你想计算它们的逐元素乘积。问题在于，这种操作对于形状不同的数组是未定义的。在这种情况下，我们可以通过以下循环来定义逐元素乘积的意义：

```
>>> out = []
>>> for i in x:
...     for j in y:
...         out.append(i*j)
...
>>> out
[0, 0, 0, 0, 0, 0, 1, 2, 3, 4, 0, 2, 4, 6, 8]
```

但现在我们失去了 x 和 y 的输入维度信息。因此，可以通过重塑输出来保持维度。

```
>>> out=np.array(out).reshape(len(x),-1)  # -1 表示推断剩余维度
>>> out
array([[0, 0, 0, 0, 0],
       [0, 1, 2, 3, 4],
       [0, 2, 4, 6, 8]])
```

关于上面的计算，另一种方式是矩阵外积。

```
>>> from numpy import matrix
>>> out=matrix(x).T * y
>>> out
matrix([[0, 0, 0, 0, 0],
        [0, 1, 2, 3, 4],
        [0, 2, 4, 6, 8]])
```

但是，如何将其推广到处理多个维度呢？让我们考虑向 y 添加一个单个维数，如下所示：

```
>>> x[:,None].shape
(3, 1)
```

为了提高可读性，可以使用 np.newaxis 替代 None。现在，如果我们直接尝试这样的操作，广播机制会通过沿着单例维度复制数据来处理不兼容的维度。

```
>>> x[:,None]*y
array([[0, 0, 0, 0, 0],
       [0, 1, 2, 3, 4],
       [0, 2, 4, 6, 8]])
```

这适用于更复杂的表达式。

```
>>> from numpy import cos
>>> x[:,None]*y + cos(x[:,None]+y)
array([[ 1.        ,  0.54030231, -0.41614684, -0.9899925 , -0.65364362],
       [ 0.54030231,  0.58385316,  1.0100075 ,  2.34635638,  4.28366219],
       [-0.41614684,  1.0100075 ,  3.34635638,  6.28366219,  8.96017029]])
```

但是，如果你不想要生成的数组的形状，该怎么办？

```
>>> x*y[:,None]  # 更改单例维度的位置
array([[0, 0, 0],
       [0, 1, 2],
       [0, 2, 4],
       [0, 3, 6],
       [0, 4, 8]])
```

现在，让我们考虑一个更大的例子。

```
>>> X = np.arange(2*4).reshape(2,4)
>>> Y = np.arange(3*5).reshape(3,5)
```

在这里，你需要按元素将这两者相乘。结果将是一个 $2 \times 4 \times 3 \times 5$ 的多维矩阵。

```
>>> X[:,:,None,None] * Y
array([[[[ 0,  0,  0,  0,  0],
         [ 0,  0,  0,  0,  0],
         [ 0,  0,  0,  0,  0]],

        [[ 0,  1,  2,  3,  4],
         [ 5,  6,  7,  8,  9],
         [10, 11, 12, 13, 14]],

        [[ 0,  2,  4,  6,  8],
         [10, 12, 14, 16, 18],
         [20, 22, 24, 26, 28]],

        [[ 0,  3,  6,  9, 12],
         [15, 18, 21, 24, 27],
         [30, 33, 36, 39, 42]]],


       [[[ 0,  4,  8, 12, 16],
         [20, 24, 28, 32, 36],
         [40, 44, 48, 52, 56]],

        [[ 0,  5, 10, 15, 20],
         [25, 30, 35, 40, 45],
         [50, 55, 60, 65, 70]],

        [[ 0,  6, 12, 18, 24],
         [30, 36, 42, 48, 54],
         [60, 66, 72, 78, 84]],

        [[ 0,  7, 14, 21, 28],
         [35, 42, 49, 56, 63],
         [70, 77, 84, 91, 98]]]])
```

第 4 章 Numpy

让我们分解这个包，看看广播对每个乘法做了什么。

```
>>> X[0,0]*Y # 第 1 个数组元素
array([[0, 0, 0, 0, 0],
       [0, 0, 0, 0, 0],
       [0, 0, 0, 0, 0]])
>>> X[0,1]*Y # 第 2 个数组元素
array([[ 0,  1,  2,  3,  4],
       [ 5,  6,  7,  8,  9],
       [10, 11, 12, 13, 14]])
>>> X[0,2]*Y # 第 3 个数组元素
array([[ 0,  2,  4,  6,  8],
       [10, 12, 14, 16, 18],
       [20, 22, 24, 26, 28]])
```

我们可以使用 axis 关键字参数沿任意维度对项求和。

```
>>> (X[:,:,None,None]*Y).sum(axis=3) # 沿第四维度求和
array([[[  0,   0,   0],
        [ 10,  35,  60],
        [ 20,  70, 120],
        [ 30, 105, 180]],

       [[ 40, 140, 240],
        [ 50, 175, 300],
        [ 60, 210, 360],
        [ 70, 245, 420]]])
```

编程技巧：使用广播缩写循环

使用广播（broadcasting）计算将 25 美分分成一分硬币（pennies）、五分硬币（nickels）和十分硬币（dimes）的方式的数量：

```
>>> n=0 # 开始计数
>>> for n_d in range(0,3): # 最多 2 个十分硬币
...     for n_n in range(0,6): # 最多 5 个五分硬币
...         for n_p in range(0,26): # 最多 25 个一分硬币
...             value = n_d*10+n_n*5+n_p
...             if value == 25:
...                 print('dimes=%d, nickels=%d, pennies=%d'%(n_d,
                                                     n_n,n_p))
...                 n+=1
...
dimes=0, nickels=0, pennies=25
dimes=0, nickels=1, pennies=20
dimes=0, nickels=2, pennies=15
dimes=0, nickels=3, pennies=10
dimes=0, nickels=4, pennies=5
dimes=0, nickels=5, pennies=0
dimes=1, nickels=0, pennies=15
dimes=1, nickels=1, pennies=10
dimes=1, nickels=2, pennies=5
dimes=1, nickels=3, pennies=0
dimes=2, nickels=0, pennies=5
dimes=2, nickels=1, pennies=0
>>> print('n = ',n)
n =  12
```

使用广播的方法可以将上述嵌套的循环简化为一行代码：

```
>>> n_d = np.arange(3)
>>> n_n = np.arange(6)
>>> n_p = np.arange(26)
>>> # 与上面的n相匹配
>>> (n_p + 5*n_n[:,None] + 10*n_d[:,None,None]==25).sum()
12
```

这意味着上面的嵌套循环和 Numpy 广播是等价的，因此，每当你看到这种嵌套循环的模式时，可能可以使用广播来简化代码。

4.13 掩码数组

Numpy 还允许屏蔽 Numpy 数组中的一部分，这在图像处理中非常常见。

```
>>> x = np.array([2, 1, 3, np.nan, 5, 2, 3, np.nan])
>>> x
array([ 2.,  1.,  3.,  nan,  5.,  2.,  3.,  nan])
>>> np.mean(x)
nan
>>> m = np.ma.masked_array(x, np.isnan(x))
>>> m
masked_array(data=[2.0, 1.0, 3.0, --, 5.0, 2.0, 3.0, --],
             mask=[False, False, False,  True, False, False,
                   False,  True],
       fill_value=1e+20)
>>> np.mean(m)
2.6666666666666665
>>> m.shape
(8,)
>>> x.shape
(8,)
>>> m.fill_value=9999
>>> m.filled()
array([2.000e+00, 1.000e+00, 3.000e+00, 9.999e+03, 5.000e+00,
       2.000e+00,
       3.000e+00, 9.999e+03])
```

编程技巧：将自定义对象转换为 Numpy 数组

为了使自定义对象与 Numpy 数组兼容，我们需要定义 __array__ 方法：

```
>>> from numpy import arange
>>> class Foo():
...     def __init__(self): # 注意双下划线
...         self.size = 10
...     def __array__(self): # numpy 数组
...         return arange(self.size)
...
>>> np.array(Foo())
array([0, 1, 2, 3, 4, 5, 6, 7, 8, 9])
```

4.14 浮点数

在计算机上用有限内存表示浮点数时存在精度限制。例如，当将两个简单数字相加时：

```
>>> 0.1 + 0.2
0.30000000000000004
```

为什么输出不是0.3？问题在这两个数字的浮点表示法以及将其相加的算法。为了用二进制表示一个整数，我们只需要用2的幂来表示。例如，230=$(11100110)_2$。Python可以使用字符串格式进行转换。

```
>>> '{0:b}'.format(230)
'11100110'
```

对于整数的加法，我们只需将相应的位相加，并将结果适配到允许的位数内。除非有溢出，否则就没有问题。表示浮点数则更为复杂，因为我们必须将这些数字表示为二进制分数。IEEE 754 标准要求浮点数表示为 $\pm C \times 2^E$，其中 C 是有效位（尾数），E 是指数。

为了将正则十进制分数表示为二进制小数，我们需要按以下形式计算分数的展开式 $a_1/2 + a_2/2^2 + a_3/2^3 \cdots$ 换句话说，需要找到 a_i 系数。可以使用与十进制小数相同的过程来实现这一点：只需除以 1/2 的分数幂，并跟踪整数部分和小数部分。Python 的 divmod 函数可以完成这方面的大部分工作。例如，将 0.125 表示为二进制分数：

```
>>> a = 0.125
>>> divmod(a*2,1)
(0.0, 0.25)
```

元组中的第一项是商，另一项是余数。如果商大于1，则相应的 a_i 项为 1，否则为 0。对于这个例子，我们有 $a_1=0$。为了得到展开式中的下一项，只需继续乘以 2，这将沿着展开式向右移动到 a_{i+1}，以此类推。然后有

```
>>> a = 0.125
>>> q,a = divmod(a*2,1)
>>> (q,a)
(0.0, 0.25)
>>> q,a = divmod(a*2,1)
>>> (q,a)
(0.0, 0.5)
>>> q,a = divmod(a*2,1)
>>> (q,a)
(1.0, 0.0)
```

当余数项为零时，该算法停止。因此，我们得到了 0.125=$(0.001)_2$。规范要求扩展中的前导项为 1。因此，有 0.125=$(1.000) \times 2^{-3}$，这意味着有效位为 1，指数为 −3。

现在，让我们回到主要问题 0.1 + 0.2，通过上述方法得到 0.1 的表示。

```
>>> a = 0.1
>>> bits = []
>>> while a>0:
...     q,a = divmod(a*2,1)
...     bits.append(q)
...
>>> ''.join(['%d'%i for i in bits])
'0001100110011001100110011001100110011001100110011001101'
```

注意，这种表示有个无限重复模式。意味着我们有 $(1.\overline{1001})_2 \times 2^{-4}$。IEEE 标准没有表示无限重复序列的方法。尽管如此，我们可以计算出：

$$\sum_{n=1}^{\infty} \frac{1}{2^{4n-3}} + \frac{1}{2^{4n}} = \frac{3}{5}$$

因此，$0.1 \approx 1.6 \times 2^{-4}$。根据 IEEE 754 标准，对于浮点型，有效位为 24 位，小数部分为 23 位。由于不能表示无限重复序列，因此必须舍入到 23 位，10011001100110011001101。因此，虽然有效位的表示是 1.6，但通过这种舍入，现在是

```
>>> b = '10011001100110011001101'
>>> 1+sum([int(i)/(2**n) for n,i in enumerate(b,1)])
1.600000023841858
```

因此，现在有 $0.1 \approx 1.600000023841858 \times 2^{-4}$=0.10000000149011612。对于 0.2 的展开，有相同的重复序列，具有不同的指数，因此有 $0.2 \approx 1.600000023841858 \times 2^{-3}$= 0.20000000298023224。要在二进制中加 0.1 + 0.2，则必须调整指数，直到它们与两者中的较高者匹配。因此，

```
  0.11001100110011001100110
+ 1.10011001100110011001101
  ------------------------
 10.01100110011001100110011
```

现在，总和必须缩小以适应有效位的可用位，因此结果是指数为 −2 的 1.00110011001100110011010。按如下所示的常规方法计算得出结果：

```
>>> k='00110011001100110011010'
>>> ('%0.12f'%((1+sum([int(i)/(2**n)
...                   for n,i in enumerate(k,1)]))/2**2)
'0.300000011921'
```

这与我们用 numpy 得到的结果相符。

```
>>> import numpy as np
>>> '%0.12f'%(np.float32(0.1) + np.float32(0.2))
'0.300000011921'
```

对于 64 位浮点数,整个过程都是相同的。Python 有分数和十进制模块,允许更精确的数字表示。十进制模块对于某些金融计算特别重要。

舍入误差

让我们考虑在 32 位浮点中 100,000,000 和 10 的求和。

```
>>> '{0:b}'.format(100000000)
'101111101011110000100000000'
```

这意味着 100,000,000=$(1.01111101011110000100000000)_2 \times 2^{26}$。同样,10=$(1.010)_2 \times 2^3$。为了求和,我们必须使指数匹配如下:

```
  1.01111101011110000100000000
+0.00000000000000000000001010
-------------------------------
  1.01111101011110000100001010
```

现在,我们必须四舍五入,因为小数点右侧只有 23 位,得到 1.01111101011110000010000,因此会丢失后面的 10 位。这使得开始时的十进制 10=$(1010)_2$ 变成了 8=$(1000)_2$。因此,再次使用 Numpy。

```
>>> format(np.float32(100000000) + np.float32(10),'10.3f')
'100000008.000'
```

这里的问题是,这两个数字之间的数量级太大,因为较小的数字右移,导致有效位的位丢失。当对这样的数字求和时,Kahan 求和算法(参见 math.fsum( ))可以有效地管理这些舍入错误。

```
>>> import math
>>> math.fsum([np.float32(100000000),np.float32(10)])
100000010.0
```

抵消误差

当两个几乎相等的浮点数相减时,会产生抵消误差(损失有效数字)。让我们考虑将 0.1111112 和 0.1111111 相减。作为二进制分数,有以下表示:

```
  1.1100011100011100100010  E-4
 -1.1100011100011100011011  E-4
 --------------------------
  0.0000000000000000001100
```

作为二元分数,这是 1.11,指数为 −23 或 $(1.75)_{10} \times 2^{-23} \approx 0.00000010430812836$。在 Numpy 中,这种精度损失如下:

```
>>> format(np.float32(0.1111112)-np.float32(0.1111111),'1.17f')
'0.00000010430812836'
```

总而言之,在使用浮点数时,必须使用诸如 Numpy 的 allclose 这样的方法来检查近似相等性,而不是通常的 Python 相等性(即==)符号。这样可以强制执行误差边界,而不是严格的相等性。同时,双精度 64 位浮点数比单精度要好得多,虽然不能消除这些问题,但可以有效地推迟它们,除非有最严格的精度要求。而 Kahan 算法对于在非常大的数据集上对浮点数进行求和而不积累舍入误差非常有效。为了最小化抵消误差，可以通过重新设计计算来避免两个几乎相等数字的相减。

> **编程技巧：十进制模块**
>
> Python 有一个内置的十进制模块，它使用不同的方法来管理浮点数，尤其是在金融计算中。缺点是十进制比这里描述的 IEEE 浮点标准慢得多。

4.15 高级 Numpy dtypes

Numpy 数据类型还可以帮助读取二进制数据文件的结构化部分。例如，WAV 文件具有 44B 的头文件格式：

```
Item                  Description
------------------    -----------------------------------
chunk_id              "RIFF"
chunk_size            4-byte unsigned little-endian integer
format                "WAVE"
fmt_id                "fmt"
fmt_size              4-byte unsigned little-endian integer
audio_fmt             2-byte unsigned little-endian integer
num_channels          2-byte unsigned little-endian integer
sample_rate           4-byte unsigned little-endian integer
byte_rate             4-byte unsigned little-endian integer
block_align           2-byte unsigned little-endian integer
bits_per_sample       2-byte unsigned little-endian integer
data_id               "data"
data_size             4-byte unsigned little-endian integer
```

你可以使用以下代码从网上打开此测试文件，以获取 chunk_id 的前四个字节。

```
>>> from urllib.request import urlopen
>>> fp=urlopen('https://www.kozco.com/tech/piano2.wav')
>>> fp.read(4)
b'RIFF'
```

继续读取接下来的四个字节得到 chunk_size。

```
>>> fp.read(4)
b'\x04z\x12\x00'
```

注意，返回的字节存在字节序（endianness）的问题。通过下面的方式获取接下来的四个字节：

```
>>> fp.read(4)
b'WAVE'
```

继续这样做，可以获得头文件的其余部分。建议使用以下自定义数据类型：

```
>>> header_dtype = np.dtype([
...     ('chunkID','S4'),
...     ('chunkSize','<u4'),
...     ('format','S4'),
...     ('subchunk1ID','S4'),
...     ('subchunk1Size','<u4'),
...     ('audioFormat','<u2'),
...     ('numChannels','<u2'),
...     ('sampleRate','<u4'),
...     ('byteRate','<u4'),
...     ('blockAlign','<u2'),
...     ('bitsPerSample','<u2'),
...     ('subchunk2ID','S4'),
...     ('subchunk2Size','u4'),
...     ])
```

另外，我们还可以使用 urlretrieve 函数来将远程数据下载到本地的临时文件。

```
>>> from urllib.request import urlretrieve
>>> path, _ = urlretrieve('https://www.kozco.com/tech/piano2.wav')
>>> h=np.fromfile(path,dtype=header_dtype,count=1)
>>> print(h)
[(b'RIFF', 1210884, b'WAVE', b'fmt ', 16, 1, 2, 48000, 192000, 4,
16, b'data', 1210848)]
```

这样我们就得到了数据的头文件信息，并能够根据头文件提供的信息来正确地对数据进行后续处理了。

> **编程技巧：格式化 Numpy 数组**
>
> 有时候，Numpy 数组可能非常密集，难以查看。np.set_printoptions 函数可以通过提供不同 Numpy 数据类型的自定义格式选项，帮助减少视觉混乱。请记住，这是一个格式化问题，不会改变底层数据。例如，要将浮点数 Numpy 数组的表示更改为 %3.2f，可以执行以下操作：
>
> `np.set_printoptions(formatter={'float':lambda i:'%3.2f'%i})`

参考文献

1. T.E. Oliphant, *A Guide to NumPy* (Trelgol Publishing, Austin, 2006)

第 5 章

Pandas

Pandas 是一个强大的模块，它在 Numpy 的基础上进行了优化，并提供了一组特别适用于分析时间序列和类电子表格数据的数据结构（可以将其想象为 Excel 中的数据透视表）。如果你熟悉 R 统计软件包，那么可以将 Pandas 看作为 Python 提供基于 Numpy 的 DataFrame。Pandas 提供了一个建立在 Numpy 平台上的 DataFrame 对象（以及其他对象），以便简化数据操作（特别是针对时间序列）用于统计处理。Pandas 在量化金融领域特别受欢迎。Pandas 的关键特性包括快速数据操作和对齐、在不同格式和 SQL 数据库之间交换数据的工具、处理缺失数据以及清理混乱数据。

5.1 使用 Series

理解 Pandas Series 对象最简单的方法是将其视作两个 Numpy 数组的容器，一个用于索引，另一个用于数据。回想一下，Numpy 数组和常规 Python 列表一样具有整数索引。

```
>>> import pandas as pd
>>> x = pd.Series([1,2,30,0,15,6])
>>> x
0     1
1     2
```

```
2    30
3     0
4    15
5     6
dtype: int64
```

该对象可以进行普通 Numpy 索引。

```
>>> x[1:3]  # Numpy 切片
1     2
2    30
dtype: int64
```

我们可以直接获得 Numpy 数组。

```
>>> x.values  # 值
array([ 1,  2, 30,  0, 15,  6])
>>> x.values[1:3]
array([ 2, 30])
>>> x.index   # 索引类似于 Numpy 数组
RangeIndex(start=0, stop=6, step=1)
```

与 Numpy 数组不同，可以使用混合类型。

```
>>> s = pd.Series([1,2,'anything','more stuff'])
>>> s
0             1
1             2
2      anything
3    more stuff
dtype: object
>>> s.index # 序列索引
RangeIndex(start=0, stop=4, step=1)
>>> s[0] # 通常的 Numpy 切片规则适用
1
>>> s[:-1]
0           1
1           2
2    anything
dtype: object
>>> s.dtype   # 对象数据类型
dtype('O')
```

注意，单个列中混合的数据类型可能会导致下游低效和其他问题。pd.Series 中的索引不仅限于整数索引。这意味着你也可以在 pd.Series 中使用字符串标签作为索引值。例如：

```
>>> s = pd.Series([1,2,3],index=['a','b','cat'])
>>> s['a']
1
>>> s['cat']
3
```

由于其作为金融数据（即股票价格）的处理工具，Pandas 非常擅长管理时间序列。

```
>>> dates = pd.date_range('20210101',periods=12)
>>> s = pd.Series(range(12),index=dates) # 显式分配索引
```

```
>>> s  # 默认为日历日
2021-01-01     0
2021-01-02     1
2021-01-03     2
2021-01-04     3
2021-01-05     4
2021-01-06     5
2021-01-07     6
2021-01-08     7
2021-01-09     8
2021-01-10     9
2021-01-11    10
2021-01-12    11
Freq: D, dtype: int64
```

你可以立刻对数据做一些基本的描述性统计（不是索引）。

```
>>> s.mean()
5.5
>>> s.std()
3.605551275463989
```

也可以使用其方法绘制（见图 5.1）序列。

```
>>> s.plot(kind='bar',alpha=0.3)  # 可以添加额外的 matplotlib 关键字
```

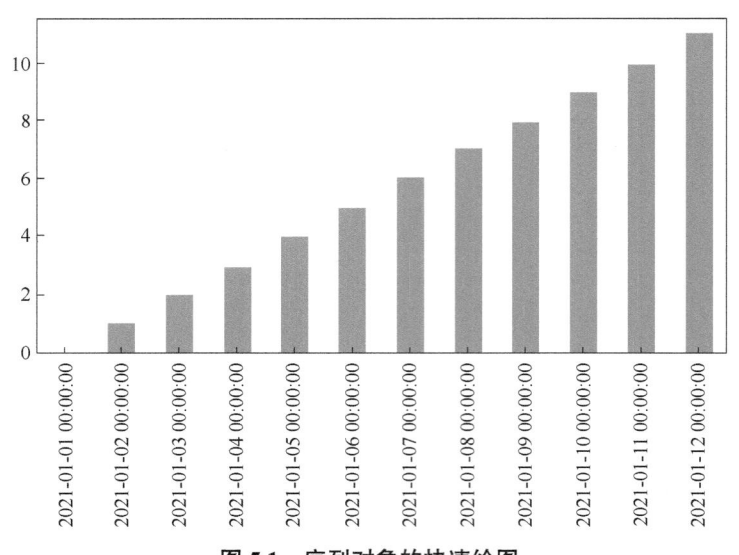

图 5.1　序列对象的快速绘图

数据可以通过索引进行汇总。例如，要计算一周中我们有数据的每一天：

```
>>> s.groupby(by=lambda i:i.dayofweek).count()
0    2
1    2
2    1
3    1
```

```
4    2
5    2
6    2
dtype: int64
```

注意，约定中 0 代表星期一，1 代表星期二，依此类推。因此，数据集中只有一个星期日（第 6 天）。groupby 方法通过 by 关键字参数给定的条件将数据划分为不相交的组。

```
>>> x = pd.Series(range(5),index=[1,2,11,9,10])
>>> x
1     0
2     1
11    2
9     3
10    4
dtype: int64
```

还可以使用模（%）运算，根据值中的元素是偶数还是奇数进行分组。

```
>>> grp = x.groupby(lambda i:i%2) # 奇数或偶数
>>> grp.get_group(0) # 偶数组
2     1
10    4
dtype: int64
>>> grp.get_group(1) # 奇数组
1     0
11    2
9     3
dtype: int64
```

在上面的代码中，第一行根据索引是偶数还是奇数将 Series 对象的元素分组。lambda 函数根据相应的索引是偶数还是奇数返回 0 或 1。接下来的一行显示了 0（即偶数）组，然后下一行显示了 1（即奇数）组。现在有了单独的分组，我们可以对每个组执行各种汇总操作，将其缩减为单个值。例如，在下面的示例中，我们得到每个组的最大值：

```
>>> grp.max() # 每组的最大值
0    4
1    3
dtype: int64
```

注意，上述操作返回的结果是另一个具有 [0，1] 元素索引的 Series 对象。分组的数量与 by 函数的唯一输出值数量相同。

5.2　使用数据帧

虽然可以将 Series 对象视为封装的两个 Numpy 数组（索引和值），但 Pandas

DataFrame 是一组共享单个索引的 Series 对象的封装。我们可以使用字典创建数据帧，如下所示：

```
>>> df = pd.DataFrame({'col1': [1,3,11,2], 'col2': [9,23,0,2]})
>>> df
   col1  col2
0     1     9
1     3    23
2    11     0
3     2     2
```

注意，输入字典中的键现在成为 DataFrame 的列标题（标签），每个对应的列与字典中对应值的列表匹配。与 Series 对象类似，DataFrame 也有一个索引，即最左边的 [0，1，2，3] 列。我们可以使用 iloc 从每列中提取元素，它忽略给定的索引并返回传统的 Numpy 切片操作。

```
>>> df.iloc[:2,:2]  # 获取部分
   col1  col2
0     1     9
1     3    23
```

也可以直接或使用"."符号对列索引，如下所示：

```
>>> df['col1']  # 索引
0     1
1     3
2    11
3     2
Name: col1, dtype: int64
>>> df.col1  # 点符号
0     1
1     3
2    11
3     2
Name: col1, dtype: int64
```

> **编程技巧：列名中的空格**
>
> 只要 DataFrame 中的列名称不包含空格或其他可评估语法（如连字符），就可以使用点符号访问列值。你也可以使用 df.columns 报告列名。

对 DataFrame 的后续操作会保留其按列结构，使用 df.sum 求每个列的总和。

```
>>> df.sum()
col1    17
col2    34
dtype: int64
```

每列的总和已经计算出来。使用 DataFrame 进行分组和聚合比 Series 更强大。

```
>>> df = pd.DataFrame({'col1': [1,1,0,0], 'col2': [1,2,3,4]})
>>> df
   col1  col2
```

```
0    1    1
1    1    2
2    0    3
3    0    4
```

在上面的 DataFrame 中，请注意 col1 列只有两个不同的值。我们可以使用以下方法对数据进行分组：

```
>>> grp=df.groupby('col1')
>>> grp.get_group(0)
   col1  col2
2    0    3
3    0    4
>>> grp.get_group(1)
   col1  col2
0    1    1
1    1    2
```

注意，每个分组对应的是 col1 中两个值之一的条目。现在，在对 col1 进行了分组后，与 Series 对象一样，也可以对每个组进行功能性汇总，如下所示：

```
>>> grp.sum()
      col2
col1
0       7
1       3
```

在每个分组中的 DataFrame 上应用求和操作。注意，上面输出的索引是原始 col1 中的每个值。

DataFrame 可以使用 eval 方法基于现有列计算新列，如下所示：

```
>>> df['sum_col']=df.eval('col1+col2')
>>> df
   col1  col2  sum_col
0    1    1      2
1    1    2      3
2    0    3      3
3    0    4      4
```

注意，你可以将输出分配给数据帧的新列。我们可以按多个列分组，如下所示：

```
>>> grp = df.groupby(['sum_col','col1'])
```

对每个组进行求和运算得到以下结果：

```
>>> res = grp.sum()
>>> res
              col2
sum_col col1
2        1      1
3        0      3
         1      2
4        0      4
```

这个输出比我们之前看到的任何内容都要复杂得多，让我们仔细地逐步分析一下。在标题下方，第一行 211 表示当 sum_col=2，col1=1 时，col2 的值为 1。

同样，对于 sum_col=3，col1=0 时，col2 的值为 3；sum_col=3，col1=1 时，col2 的值为 2。这种分层显示是查看结果的一种方式。注意，上面的层次不是均匀的。我们可以对此结果进行 unstack 操作。

```
>>> res.unstack()
        col2
col1       0    1
sum_col
2        NaN  1.0
3        3.0  2.0
4        4.0  NaN
```

NaN 值表示表格中此位置为空。例如，对于（sum_col=2，col2=0）这对组合，DataFrame 中没有相应的值，可以通过查看倒数第二个代码块进行验证。同样，对于（sum_col=4，col2=1）这一对也没有相对应的值。因此，这表明倒数第二个代码块中的原始表示与这里的表示是相同的，只是这里明确显示了上述由 NaN 表示的缺失值。

让我们继续索引数据帧：

```
>>> import numpy as np
>>> data=np.arange(len(dates)*4).reshape(-1,4)
>>> df = pd.DataFrame(data,index=dates,
...                   columns=['A','B','C','D' ])
>>> df
             A   B   C   D
2021-01-01   0   1   2   3
2021-01-02   4   5   6   7
2021-01-03   8   9  10  11
2021-01-04  12  13  14  15
2021-01-05  16  17  18  19
2021-01-06  20  21  22  23
2021-01-07  24  25  26  27
2021-01-08  28  29  30  31
2021-01-09  32  33  34  35
2021-01-10  36  37  38  39
2021-01-11  40  41  42  43
2021-01-12  44  45  46  47
```

现在，可以按名称访问每个列，如下所示：

```
>>> df['A']
2021-01-01     0
2021-01-02     4
2021-01-03     8
2021-01-04    12
2021-01-05    16
2021-01-06    20
2021-01-07    24
2021-01-08    28
2021-01-09    32
2021-01-10    36
2021-01-11    40
2021-01-12    44
Freq: D, Name: A, dtype: int64
```

或者，使用更快的属性样式表示法。

```
>>> df.A
2021-01-01     0
2021-01-02     4
2021-01-03     8
2021-01-04    12
2021-01-05    16
2021-01-06    20
2021-01-07    24
2021-01-08    28
2021-01-09    32
2021-01-10    36
2021-01-11    40
2021-01-12    44
Freq: D, Name: A, dtype: int64
```

我们可以做一些基本的计算和索引。

```
>>> df.loc[:dates[3]] # 与Python的惯例不同,这里包含端点
             A   B   C   D
2021-01-01   0   1   2   3
2021-01-02   4   5   6   7
2021-01-03   8   9  10  11
2021-01-04  12  13  14  15
>>> df.loc[:,'A':'C'] # 通过列标签切片选择所有行
             A   B   C
2021-01-01   0   1   2
2021-01-02   4   5   6
2021-01-03   8   9  10
2021-01-04  12  13  14
2021-01-05  16  17  18
2021-01-06  20  21  22
2021-01-07  24  25  26
2021-01-08  28  29  30
2021-01-09  32  33  34
2021-01-10  36  37  38
2021-01-11  40  41  42
2021-01-12  44  45  46
```

5.3 重新索引

DataFrame 或 Series 具有可以用来对齐数据的重新索引。

```
>>> x = pd.Series(range(3),index=['a','b','c'])
>>> x
a    0
b    1
c    2
dtype: int64
>>> x.reindex(['c','b','a','z'])
c    2.0
b    1.0
a    0.0
z    NaN
dtype: float64
```

注意，新创建的 Series 对象具有新索引，并使用 NaN 填充缺失的项。你可以通过在 reindex 中使用 fill_value 关键字参数来填充其他值。在处理有序数据时，也可以进行向后填充和向前填充（ffill）值，如下所示：

```
>>> x = pd.Series(['a','b','c'],index=[0,5,10])
>>> x
0     a
5     b
10    c
dtype: object
>>> x.reindex(range(11),method='ffill')
0     a
1     a
2     a
3     a
4     a
5     b
6     b
7     b
8     b
9     b
10    c
dtype: object
```

更复杂的插值方法也是可实现的，但不能直接使用 reindex。重新索引也适用于 DataFrame，可以实现在两个维度中的一个或两个上进行重新索引。

```
>>> df = pd.DataFrame(index=['a','b','c'],
...                   columns=['A','B','C','D'],
...                   data = np.arange(3*4).reshape(3,4))
>>> df
   A  B   C   D
a  0  1   2   3
b  4  5   6   7
c  8  9  10  11
```

现在，我们可以通过下面的方法对其重新索引：

```
>>> df.reindex(['c','b','a','z'])
     A    B     C     D
c  8.0  9.0  10.0  11.0
b  4.0  5.0   6.0   7.0
a  0.0  1.0   2.0   3.0
z  NaN  NaN   NaN   NaN
```

注意，缺失的 z 行是如何被填充的。相同的方法也适用于对列重新索引，如下所示：

```
>>> df.reindex(columns=['D','A','C','Z','B'])
    D  A   C    Z  B
a   3  0   2  NaN  1
b   7  4   6  NaN  5
c  11  8  10  NaN  9
```

z 列被自动填充了 NaN。同样，反向填充和正向填充也可以实现。

5.4 删除项目

以 Series 对象为例：
```
>>> x = pd.Series(range(3),index=['a','b','c'])
```
为了去掉 "a" 索引处的数据，我们可以使用 drop 方法，该方法将返回一个新的 Series 对象，并删除指定的数据。
```
>>> x.drop('a')
b    1
c    2
dtype: int64
```
注意，这是一个新的 Series 对象，除非我们使用 inplace 关键字参数或显式地使用 del。
```
>>> del x['a']
```
此删除方法同样适用于 DataFrame。
```
>>> df = pd.DataFrame(index=['a','b','c'],
...                   columns=['A','B','C','D'],
...                   data = np.arange(3*4).reshape(3,4))
>>> df.drop('a')
   A  B   C   D
b  4  5   6   7
c  8  9  10  11
```
或者，沿着列方向。
```
>>> df.drop('A',axis=1)
   B   C   D
a  1   2   3
b  5   6   7
c  9  10  11
```
同样，del 和 inplace 也适用于 DataFrame。

5.5 高级索引

Pandas 提供了非常强大和快速的切片（slicing）功能。
```
>>> x = pd.Series(range(4),index=['a','b','c','d'])
>>> x['a':'c']
a    0
b    1
c    2
dtype: int64
```

注意，与常规 Python 索引不同，在使用标签进行切片时两个端点都是包含在内的。高级索引还可以用来赋值。

```
>>> x['a':'c']=999
>>> x
a    999
b    999
c    999
d      3
dtype: int64
```

类似的操作也适用于 DataFrame。

```
>>> df = pd.DataFrame(index=['a','b','c'],
...                   columns=['A','B','C','D'],
...                   data = np.arange(3*4).reshape(3,4))
>>> df['a':'b']
   A  B  C  D
a  0  1  2  3
b  4  5  6  7
```

你还可以选择单个列。

```
>>> df[['A','C']]
   A   C
a  0   2
b  4   6
c  8  10
```

可以使用 loc 将基于标签的切片与类似于 Numpy 的冒号切片混合使用。

```
>>> df.loc['a':'b',['A','C']]
   A  C
a  0  2
b  4  6
```

其中，第一个参数用于索引行，第二个参数用于索引列。这种方法允许使用类似于 Numpy 的索引，而不必担心标签。如果要返回到普通的 Numpy 索引，可以使用 iloc。

```
>>> df.iloc[0,-2:]
C    2
D    3
Name: a, dtype: int64
```

5.6 广播和数据对齐

在操作一个或多个 Series 或 DataFrame 对象时，要记住的主要一点是，索引始终会对齐计算。

```
>>> x = pd.Series(range(4),index=['a','b','c','d'])
>>> y = pd.Series(range(3),index=['a','b','c'])
```

注意，y 相比 x 缺少一个索引，所以求和时，
```
>>> x+y
a    0.0
b    2.0
c    4.0
d    NaN
dtype: float64
```

y 缺少的索引由 NaN 填充了。这种操作同样适用于 DataFrame。
```
>>> df = pd.DataFrame(index=['a','b','c'],
...                   columns=['A','B','C','D'],
...                   data = np.arange(3*4).reshape(3,4))
>>> ef = pd.DataFrame(index=list('abcd'),
...                   columns=list('ABCDE'),
...                   data = np.arange(4*5).reshape(4,5))
>>> ef
    A   B   C   D   E
a   0   1   2   3   4
b   5   6   7   8   9
c  10  11  12  13  14
d  15  16  17  18  19
>>> df
   A  B   C   D
a  0  1   2   3
b  4  5   6   7
c  8  9  10  11
>>> df+ef
      A     B     C     D    E
a   0.0   2.0   4.0   6.0  NaN
b   9.0  11.0  13.0  15.0  NaN
c  18.0  20.0  22.0  24.0  NaN
d   NaN   NaN   NaN   NaN  NaN
```

注意，非重叠元素将用 NaN 填充。你也可以使用命名操作来指定填充值，例如：
```
>>> df.add(ef,fill_value=0)
      A     B     C     D     E
a   0.0   2.0   4.0   6.0   4.0
b   9.0  11.0  13.0  15.0   9.0
c  18.0  20.0  22.0  24.0  14.0
d  15.0  16.0  17.0  18.0  19.0

>>> s = df.loc['a']  # 第一行
>>> s
A    0
B    1
C    2
D    3
Name: a, dtype: int64
```

进行求和时，我们将获得以下结果：
```
>>> s + df
   A   B   C   D
a  0   2   4   6
b  4   6   8  10
c  8  10  12  14
```

将其与原始 DataFrame 进行比较。

```
>>> df
   A  B  C   D
a  0  1  2   3
b  4  5  6   7
c  8  9  10  11
```

这表明 Series 对象被广播到行中，与 DataFrame 中的列对齐。下面是一个不同的 Series 对象的示例，该对象缺少 DataFrame 中的某些列。

```
>>> s = pd.Series([1,2,3],index=['A','D','E'])
>>> s+df
     A    B    C     D   E
a  1.0  NaN  NaN   5.0  NaN
b  5.0  NaN  NaN   9.0  NaN
c  9.0  NaN  NaN  13.0  NaN
```

注意，广播仍然沿着行进行，与列对齐，但会用 NaN 填充缺失的条目。

这里有一个快速的 Python 测试，它使用正则表达式来测试相对较小的质数。

```
>>> import re
>>> pattern = r'^1?$|^(11+?)\1+$'
>>> def isprime(n):
...     return (re.match(pattern, '1'*n) is None) # *
...
```

现在，我们可以找到哪个列标签中质数最多。

```
>>> df.applymap(isprime)
       A      B      C     D
a  False  False   True  True
b  False   True  False  True
c  False  False  False  True
```

布尔值自动转换。

```
>>> df.applymap(isprime).sum()
A    0
B    1
C    1
D    3
dtype: int64
```

> **编程技巧：Pandas 性能**
>
> Pandas 的 groupby、apply 和 applymap 方法非常灵活和强大，但是 Pandas 必须在 Python 解释器中进行评估，而不是在优化的 Pandas 代码中进行评估，这会导致显著的减速。因此，最好始终使用内置于 Pandas 中的函数，而不是定义纯 Python 函数来传递给这些方法。

5.7 分类和合并

Pandas 支持一些关系代数运算，如表连接。
```
>>> df = pd.DataFrame(index=['a','b','c'],
...                   columns=['A','B','C'],
...                   data = np.arange(3*3).reshape(3,3))
>>> df
   A  B  C
a  0  1  2
b  3  4  5
c  6  7  8
>>> ef = pd.DataFrame(index=['a','b','c'],
...                   columns=['A','Y','Z'],
...                   data = np.arange(3*3).reshape(3,3))
>>> ef
   A  Y  Z
a  0  1  2
b  3  4  5
c  6  7  8
```
表连接由 merge 函数实现。
```
>>> pd.merge(df,ef,on='A')
   A  B  C  Y  Z
0  0  1  2  1  2
1  3  4  5  4  5
2  6  7  8  7  8
```
关键字参数 on 表示在具有 A 列中匹配对应条目的情况下合并这两个 DataFrame。注意，在合并中索引并未保留。为了让例子更有趣，我们通过删除一行来修改 ef DataFrame。
```
>>> ef.drop('b',inplace=True)
>>> ef
   A  Y  Z
a  0  1  2
c  6  7  8
```
将 ef 删除一行后，让我们再次尝试合并，看看会发生什么。
```
>>> pd.merge(df,ef,on='A')
   A  B  C  Y  Z
0  0  1  2  1  2
1  6  7  8  7  8
```
注意，合并后仅保留与两个 DataFrame 都匹配（即在两者的交集中）的 A 元素。我们可以通过使用 how 关键字参数来改变这一点。
```
>>> pd.merge(df,ef,on='A',how='left')
   A  B  C    Y    Z
0  0  1  2  1.0  2.0
1  3  4  5  NaN  NaN
2  6  7  8  7.0  8.0
```

关键字参数 how=left 告诉连接保留左侧 DataFrame（在这种情况下是 df）上的所有键，并在右侧 DataFrame（ef）中缺失的地方填充 NaN。如果 ef 中存在 df 中没有的 A 元素，则这些元素将消失。

```
>>> ef = pd.DataFrame(index=['a','d','c'],
...                   columns=['A','Y','Z'],
...                   data = 10*np.arange(3*3).reshape(3,3))
>>> ef
   A   Y   Z
a  0  10  20
d 30  40  50
c 60  70  80
>>> pd.merge(df,ef,on='A',how='left')
   A  B  C     Y     Z
0  0  1  2  10.0  20.0
1  3  4  5   NaN   NaN
2  6  7  8   NaN   NaN
```

同样，我们可以使用 how=right 关键字。

```
>>> pd.merge(df,ef,on='A',how='right')
    A    B    C   Y   Z
0   0  1.0  2.0  10  20
1  30  NaN  NaN  40  50
2  60  NaN  NaN  70  80
```

我们还可以使用 how=outer 关键字来获得并集。

```
>>> pd.merge(df,ef,on='A',how='outer')
    A    B    C     Y     Z
0   0  1.0  2.0  10.0  20.0
1   3  4.0  5.0   NaN   NaN
2   6  7.0  8.0   NaN   NaN
3  30  NaN  NaN  40.0  50.0
4  60  NaN  NaN  70.0  80.0
```

另一个常见的任务是将连续数据拆分为离散。

```
>>> a = np.arange(10)
>>> a
array([0, 1, 2, 3, 4, 5, 6, 7, 8, 9])
>>> bins = [0,5,10]
>>> cats = pd.cut(a,bins)
>>> cats
[NaN, (0.0, 5.0], (0.0, 5.0], (0.0, 5.0], (0.0, 5.0], (0.0, 5.0],
(5.0, 10.0], (5.0, 10.0], (5.0, 10.0], (5.0, 10.0]]
Categories (2, interval[int64]): [(0, 5] < (5, 10]]
```

pd.cut 函数获取数组中的数据，并将其放入分类变量 cats 中。

```
>>> cats.categories
IntervalIndex([(0, 5], (5, 10]],
              closed='right',
              dtype='interval[int64]')
```

半开区间表示每个类别的边界。你可以通过传递 right=False 关键字参数来更改区间的奇偶性。

```
>>> cats.codes
array([-1, 0, 0, 0, 0, 0, 1, 1, 1, 1], dtype=int8)
```

上面的 –1 意味着 0 不包含在这两个类别中，因为区间在左侧是开放的。你可以按如下所示统计每个类别中的元素数量：

```
>>> pd.value_counts(cats)
(0, 5]     5
(5, 10]    4
dtype: int64
```

可以使用 labels 关键字参数传递每个类别的描述性名称。

```
>>> cats = pd.cut(a,bins,labels=['one','two'])
>>> cats
[NaN, 'one', 'one', 'one', 'one', 'one', 'two', 'two', 'two', 'two']
Categories (2, object): ['one' < 'two']
```

注意，如果为 bins 参数传递一个整数参数，它将自动分割为大小相等的类别。qcut 函数非常类似，只是它是根据四分位数进行分割的。

```
>>> a = np.random.rand(100)  # 均匀随机变量
>>> cats = pd.qcut(a,4,labels=['q1','q2','q3','q4'])
>>> pd.value_counts(cats)
q4    25
q3    25
q2    25
q1    25
dtype: int64
```

5.8 内存使用和数据类型 dtypes

在实际的开发工作中，你可能会使用 Pandas 来处理大量数据。以下是一些有效处理数据的技巧。首先，我们需要使用 Seaborn 中的 Penguins 数据集。

```
>>> import seaborn as sns
>>> df = sns.load_dataset('penguins')
>>> df.head()
  species     island  bill_length_mm  bill_depth_mm  flipper_length_mm  body_mass_g     sex
0  Adelie  Torgersen            39.1           18.7              181.0       3750.0    Male
1  Adelie  Torgersen            39.5           17.4              186.0       3800.0  Female
2  Adelie  Torgersen            40.3           18.0              195.0       3250.0  Female
3  Adelie  Torgersen             NaN            NaN                NaN          NaN     NaN
4  Adelie  Torgersen            36.7           19.3              193.0       3450.0  Female
```

这不是一个特别大的数据集，但它已经足够满足我们的需要。首先，让我们检查一下数据帧的数据类型。

```
>>> df.dtypes
species            object
island             object
bill_length_mm    float64
bill_depth_mm     float64
```

```
flipper_length_mm       float64
body_mass_g             float64
sex                      object
dtype: object
```

注意，上面有一些标记为 object。这通常意味着效率低下，因为这种通用数据类型可能会消耗过多的内存。Pandas 提供了一种评估数据帧内存消耗的简单方法。

```
>>> df.memory_usage(deep=True)
Index                128
species            21876
island             21704
bill_length_mm      2752
bill_depth_mm       2752
flipper_length_mm   2752
body_mass_g         2752
sex                20995
dtype: int64
```

现在，我们对这个数据帧的内存消耗有所了解，可以通过更改数据类型来改进它。我们可以指定 sex 列的 dtype 为前面讨论过的分类类型。

```
>>> ef = df.astype({'sex':'category'})
>>> ef.memory_usage(deep=True)
Index                128
species            21876
island             21704
bill_length_mm      2752
bill_depth_mm       2752
flipper_length_mm   2752
body_mass_g         2752
sex                  548
dtype: int64
```

数据类型的修改将 sex 的内存消耗减少了将近 40 倍，如果 DataFrame 有数千行的话，这将是非常显著的。让我们继续使用 category 作为 species 和 island 列的数据类型。

```
>>> ef = df.astype({'sex':'category',
...                 'species':'category',
...                 'island':'category'})
>>> ef.memory_usage(deep=True)
Index                128
species              616
island               615
bill_length_mm      2752
bill_depth_mm       2752
flipper_length_mm   2752
body_mass_g         2752
sex                  548
dtype: int64
```

为了进行比较，我们可以将这些放入一个新的数据帧中。

```
>>> (pd.DataFrame({'df':df.memory_usage(deep=True),
...                'ef':ef.memory_usage(deep=True)})
...   .assign(ratio= lambda i:i.ef/i.df))
                       df     ef     ratio
Index                 128    128  1.000000
species             21876    616  0.028159
island              21704    615  0.028336
bill_length_mm       2752   2752  1.000000
bill_depth_mm        2752   2752  1.000000
flipper_length_mm    2752   2752  1.000000
body_mass_g          2752   2752  1.000000
sex                 20995    548  0.026101
```

上面显示了我们将 dtype 更改后列的内存占用空间要小得多。如果我们不需要 float64 默认的精度水平，也可以将数值类型进行更改。例如，flipper_length_mm 列是以毫米为单位测量的，且所有测量值都没有小数部分。因此，我们可以将该列更改为以下数据类型，从而节省四倍的内存。

```
>>> ef = df.astype({'sex':'category',
...                 'species':'category',
...                 'island':'category',
...                 'flipper_length_mm': np.float16})
>>> ef.memory_usage(deep=True)
Index                 128
species               616
island                615
bill_length_mm       2752
bill_depth_mm        2752
flipper_length_mm     688
body_mass_g          2752
sex                   548
dtype: int64
```

以下是再次总结：

```
>>> (pd.DataFrame({'df':df.memory_usage(deep=True),
...                'ef':ef.memory_usage(deep=True)})
...   .assign(ratio= lambda i:i.ef/i.df))
                       df     ef     ratio
Index                 128    128  1.000000
species             21876    616  0.028159
island              21704    615  0.028336
bill_length_mm       2752   2752  1.000000
bill_depth_mm        2752   2752  1.000000
flipper_length_mm    2752    688  0.250000
body_mass_g          2752   2752  1.000000
sex                 20995    548  0.026101
```

因此，将默认的 object 数据类型更改为其他更小的数据类型可以显著地节省内存，并可能加快数据框的下游处理速度。当直接从网络上提取数据到 Dataframe 时，尤其是使用 pd.read_html，即使网页上的数据看起来是数值型的，通常也会存在不必要的 object 数据类型。

5.9 常见的操作

一个常见的问题是如何将字符串列拆分为组件。例如:
```
>>> df = pd.DataFrame(dict(name=['Jon Doe','Jane Smith']))
>>> df.name.str.split(' ',expand=True)
      0      1
0   Jon    Doe
1  Jane  Smith
```
关键步骤是 expand 关键字参数,它将结果转换为 DataFrame。该结果可以使用下面的方法分配到同一 DataFrame 中:
```
>>> df[['first','last']]=df.name.str.split(' ',expand=True)
>>> df
         name  first   last
0     Jon Doe    Jon    Doe
1  Jane Smith   Jane  Smith
```
注意,如果不使用 expand 关键字,将导致输出为列表而不是 DataFrame。
```
>>> df.name.str.split(' ')
0      [Jon, Doe]
1    [Jane, Smith]
Name: name, dtype: object
```
这可以通过在输出上使用 apply 将 list 转换为 Series 对象来解决。
```
>>> df.name.str.split(' ').apply(pd.Series)
      0      1
0   Jon    Doe
1  Jane  Smith
```

apply 方法是最强大且通用的 DataFrame 方法之一。它作用于 DataFrame 的各个列(即 pd.Series 对象)。与 applymap 方法不同,apply 方法可以返回不同形状的对象。例如,在具有数值行的 DataFrame 上执行类似 df.apply(lambda i: i[i>3])的操作将返回一个小型截断的 NaN 填充的 DataFrame。此外,通过在 df.apply(raw=True)关键字参数中直接在 DataFrame 列的底层 Numpy 数组上操作,可以加快方法的速度。这意味着 'apply' 方法直接处理 Numpy 数组,而不是通常的 'pd.Series' 对象。

transform 方法与 apply 密切相关,但必须生成具有相同维度的输出 DataFrame,例如:
```
>>> df = pd.DataFrame({'A': [1,1,2,2], 'B': range(4)})
>>> df
   A  B
0  1  0
```

```
1  1  1
2  2  2
3  2  3
```

我们可以按"A"分组。

```
>>> df.groupby('A').sum()
   B
A
1  1
2  5
```

但是通过使用 .transform( )，我们可以将结果广播到原始 DataFrame 中各自的位置。

```
>>> df.groupby('A').transform('sum')
   B
0  1
1  1
2  5
3  5
```

describe 方法可用于总结给定的 DataFrame，如下所示：

```
>>> df = pd.DataFrame(index=['a','b','c'], columns=['A','B','C'],
↪    data = np.arange(3*3).reshape(3,3))
>>> df.describe()
         A    B    C
count  3.0  3.0  3.0
mean   3.0  4.0  5.0
std    3.0  3.0  3.0
min    0.0  1.0  2.0
25%    1.5  2.5  3.5
50%    3.0  4.0  5.0
75%    4.5  5.5  6.5
max    6.0  7.0  8.0
```

删除意外出现的重复条目通常也很有用。

```
>>> df = pd.DataFrame({'A': [1,1,2,2,2,3], 'B': range(6)})
>>> df.drop_duplicates('A')
   A  B
0  1  0
2  2  2
5  3  5
```

keep 关键字参数可以决定要保留哪个重复的条目。

5.10 显示 DataFrame

Pandas 提供了 set_option 方法来更改 DataFrame 的可视显示，同时不改变相应的数据。

```
>>> pd.set_option('display.float_format','{:.2f}'.format)
```

注意，参数是一个可调用对象，用于生成格式化的字符串。这些自定义设置可以通过 reset_option 恢复，例如：

```
pd.reset_option('display.float_format')
```

chop 选项非常方便，可以用于修剪过多的显示精度，例如：

```
>>> pd.set_option('display.chop',1e-5)
```

在 Jupyter Notebook 中，格式化可以利用 HTML 元素与 style.format DataFrame 方法，如图 5.2 所示。

```
>>> from pandas_datareader import data
>>> df=data.DataReader("F", 'yahoo', '20200101',
↪    '20200110').reset_index()
>>> (df.style.format(dict(Date='{:%m/%d/%Y}'))
...   .hide_index()
...   .highlight_min('Close',color='red')
...   .highlight_max('Close',color='lightgreen')
... )
<pandas.io.formats.style.Styler object at 0x7f9376a22460>
```

日期	最高价	最低价	开盘价	收盘价	交易量	调整后收盘价
01/02/2020	9.420000	9.190000	9.290000	9.420000	43425700	9.262475
01/03/2020	9.370000	9.150000	9.310000	9.210000	45040800	9.055987
01/06/2020	9.170000	9.060000	9.100000	9.160000	43372300	9.006823
01/07/2020	9.250000	9.120000	9.200000	9.250000	44984100	9.095318
01/08/2020	9.300000	9.170000	9.230000	9.250000	45994900	9.095318
01/09/2020	9.310000	9.180000	9.300000	9.260000	51817400	9.105151
01/10/2020	9.360000	9.250000	9.270000	9.250000	39796300	9.095318

图 5.2　DataFrame 中的高亮显示项目

生成的表格中的最低收盘价用深灰色突出显示，最高收盘价用浅灰色突出显示。提供这种快速视觉提示对于提取关键数据元素至关重要。注意，上面的括号是为了使用换行符分隔点方法，这是从 R DataFrame 继承的一种风格。关键步骤是使用 format() 暴露样式格式，然后使用其方法在 Jupyter Notebook 中样式化生成的 HTML 表格。下面的代码更改了 Volume 列的颜色渐变，如图 5.3 所示。

```
>>> (df.style.format(dict(Date='{:%m/%d/%Y}'))
...   .hide_index()
...   .background_gradient(subset='Volume',cmap='Blues')
... )
<pandas.io.formats.style.Styler object at 0x7f9376a8a0d0>
```

在 Jupyter Notebook 的表格表示中，还可以嵌入背景柱状图，代码示例如下

（见图 5.4）：

```
>>> (df.style.format(dict(Date='{:%m/%d/%Y}'))
...   .hide_index()
...   .bar('Volume',color='lightblue',align='zero')
... )
<pandas.io.formats.style.Styler object at 0x7f9374711f70>
```

日期	最高价	最低价	开盘价	收盘价	交易量	调整后收盘价
01/02/2020	9.420000	9.190000	9.290000	9.420000	43425700	9.262475
01/03/2020	9.370000	9.150000	9.310000	9.210000	45040800	9.055987
01/06/2020	9.170000	9.060000	9.100000	9.160000	43372300	9.006823
01/07/2020	9.250000	9.120000	9.200000	9.250000	44984100	9.095318
01/08/2020	9.300000	9.170000	9.230000	9.250000	45994900	9.095318
01/09/2020	9.310000	9.180000	9.300000	9.260000	51817400	9.105151
01/10/2020	9.360000	9.250000	9.270000	9.250000	39796300	9.095318

图 5.3　用于 HTML DataFrame 渲染的自定义颜色梯度

日期	最高价	最低价	开盘价	收盘价	交易量	调整后收盘价
01/02/2020	9.420000	9.190000	9.290000	9.420000	43425700	9.262475
01/03/2020	9.370000	9.150000	9.310000	9.210000	45040800	9.055987
01/06/2020	9.170000	9.060000	9.100000	9.160000	43372300	9.006823
01/07/2020	9.250000	9.120000	9.200000	9.250000	44984100	9.095318
01/08/2020	9.300000	9.170000	9.230000	9.250000	45994900	9.095318
01/09/2020	9.310000	9.180000	9.300000	9.260000	51817400	9.105151
01/10/2020	9.360000	9.250000	9.270000	9.250000	39796300	9.095318

图 5.4　用于 DataFrame 的自定义背景柱状图

5.11　分层索引

当我们在使用多个列进行分组时，会遇到需要使用 Pandas 的 MultiIndex 的情况。

```
>>> idx = pd.MultiIndex.from_product([['a','b'],[1,2,3]])
>>> data = 10*np.arange(6).reshape(6,1)
>>> df = pd.DataFrame(data=data,index=idx,columns=['A'])
>>> df
      A
a 1   0
  2  10
  3  20
```

```
b 1    30
  2    40
  3    50
```

它比以下内容更紧凑。

```
>>> df.reset_index()
  level_0  level_1   A
0    a        1      0
1    a        2     10
2    a        3     20
3    b        1     30
4    b        2     40
5    b        3     50
```

由于我们创建索引时没有为其命名，因此表头是 level_0 和 level_1。我们可以在 DataFrame 中交换索引的两个级别。

```
>>> df.swaplevel()
         A
1 a      0
2 a     10
3 a     20
1 b     30
2 b     40
3 b     50
```

pd.IndexSlice 的使用让 loc 对 DataFrame 的索引变得更加容易。

```
>>> ixs = pd.IndexSlice
>>> df.loc[ixs['a',:],:]
       A
a 1    0
  2   10
  3   20
>>> df.loc[ixs[:,2],:]
       A
a 2   10
b 2   40
```

注意，分层索引可以有更多层级。上面的操作也适用于列索引。

```
>>> rx = pd.MultiIndex.from_product([['a','b'],[1,2,3]])
>>> cx = pd.MultiIndex.from_product([['A','B','C'],[2,3]])
>>> data=[[2, 3, 9, 3, 4, 1],
...       [9, 5, 9, 7, 2, 1],
...       [9, 4, 4, 3, 2, 1],
...       [1, 0, 4, 5, 5, 5],
...       [5, 8, 1, 6, 1, 7],
...       [0, 8, 9, 2, 1, 9]]
>>> df = pd.DataFrame(index=rx,columns=cx,data=data)
>>> df
       A     B     C
       2 3   2 3   2 3
a 1    2 3   9 3   4 1
  2    9 5   9 7   2 1
  3    9 4   4 3   2 1
b 1    1 0   4 5   5 5
  2    5 8   1 6   1 7
  3    0 8   9 2   1 9
```

也可以对列和行使用 pd.IndexSlice。
```
>>> df.loc[ixs['a',:],ixs['A',:]]=1
>>> df
     A     B     C
     2  3  2  3  2  3
a 1  1  1  9  3  4  1
  2  1  1  9  7  2  1
  3  1  1  4  3  2  1
b 1  1  0  4  5  5  5
  2  5  8  1  6  1  7
  3  0  8  9  2  1  9
```
将名称添加到索引级别很有帮助。
```
>>> df.index = df.index.set_names(['X','Y'])
>>> df
       A     B     C
       2  3  2  3  2  3
X Y
a 1    1  1  9  3  4  1
  2    1  1  9  7  2  1
  3    1  1  4  3  2  1
b 1    1  0  4  5  5  5
  2    5  8  1  6  1  7
  3    0  8  9  2  1  9
```
即使行/列上有这些复杂的多层索引，groupby 方法仍然有效，但需要完全指定特定的列，例如 ('B', 2)。
```
>>> df.groupby(('B',2)).sum()
        A        B        C
        2   3    2   3    2    3
(B, 2)
1       5   8    6   1    7
4       2   1    8   7    6
9       2  10   12   7   11
```
为了理解这是如何工作的，需要取出列切片并检查其元素。
```
>>> df.loc[:,('B',2)].unique()
array([9, 4, 1])
```
现在，我们检查 DataFrame 中由这些值创建的分区，例如：
```
>>> df.groupby(('B',2)).get_group(4)
       A     B     C
       2  3  2  3  2  3
X Y
a 3    1  1  4  3  2  1
b 1    1  0  4  5  5  5
```
然后对这些分组求和产生最终输出。你还可以在组上使用 apply 函数来计算非标量输出。例如，要减去组中每个元素的最小值，我们可以执行以下操作：
```
>>> df.groupby(('B',2)).apply(lambda i:i-i.min())
       A     B     C
       2  3  2  3  2  3
X Y
a 1    1  0  0  1  3  0
```

```
  2  1  0  0  5  1  0
  3  0  1  0  0  0  0
b 1  0  0  0  2  3  4
  2  0  0  0  0  0  0
  3  0  7  0  0  0  8
```

5.12 Pipes

Pandas 使用 pipe 函数实现方法链接。尽管这不符合 Python 的风格，但比起将操作 DataFrame 的函数逐个组合在一起更容易。

```
>>> df = pd.DataFrame(index=['a','b','c'],
...                   columns=['A','B','C'],
...                   data = np.arange(3*3).reshape(3,3))
>>> df.pipe(lambda i:i*10).pipe(lambda i:3*i)
     A    B    C
a    0   30   60
b   90  120  150
c  180  210  240
```

假设我们需要找出列求和为奇数的情况。可以使用 assign 创建一个临时变量 t，然后从 DataFrame 中提取相应部分，如下所示：

```
>>> df.assign(t=df.A+df.B+df.C).query('t%2==1').drop('t',axis=1)
   A  B  C
a  0  1  2
c  6  7  8
```

assign 方法接收一个函数，其参数可以是 DataFrame 本身或命名的 DataFrame。然后 query 方法根据 t 的奇偶性过滤中间结果，最后一步是删除输出中不再需要的 t 变量。

5.13 数据文件和数据库

Pandas 具有强大的 I/O 工具，可用于操作 Excel 和 CSV 电子表格。

```
>>> df.to_excel('this_excel.file.xls')
```

给定的电子表格的日期格式符合 Excel 的内部日期表示。

如果安装了 PyTables，则可以写入 HDFStore。你还可以直接从 PyTables 操作 HDF5。

```
>>> df.to_hdf('filename.h5','keyname')
```

可以使用以下方式获取数据。

```
>>> dg=pd.read_hdf('filename.h5','keyname')
```

可以创建 SQLite 数据库，因为 SQLite 包含在 Python 中。

```
>>> import sqlite3
>>> cnx = sqlite3.connect(':memory:')
>>> df = pd.DataFrame(index=['a','b','c'],
...                   columns=['A','B','C'],
...                   data = np.arange(3*3).reshape(3,3))
>>> df.to_sql('TableName',cnx)
```

可以从数据库重新加载。

```
>>> from pandas.io import sql
>>> p2 = sql.read_sql_query('select * from TableName', cnx)
>>> p2
  index  A  B  C
0     a  0  1  2
1     b  3  4  5
2     c  6  7  8
```

5.14 自定义 Pandas

自从 Pandas 0.23 版本以来，可以使用 extensions.register_dataframe_accessor 来扩展 Pandas 的 DataFrame/Series，而无需进行子类化，这样更加方便。

```
>>> df = pd.DataFrame(index=['a','b','c'],
...                   columns=['A','B','C','D'],
...                   data = np.arange(3*4).reshape(3,4))
>>> df
   A  B   C   D
a  0  1   2   3
b  4  5   6   7
c  8  9  10  11
```

以下代码定义了一个自定义访问器，其行为就像是一个原生的 DataFrame 方法一样。

```
>>> @pd.api.extensions.register_dataframe_accessor('custom')
... class CustomAccess:
...     def __init__(self,df): # 接收 DataFrame
...         assert 'A' in df.columns # 一些输入验证
...         assert 'B' in df.columns
...         self._df = df
...     @property   # 自定义属性
...     def odds(self):
...         'drop all columns that have all even elements'
...         df = self._df
...         return df[df % 2==0].dropna(axis=1,how='all')
...     def avg_odds(self): # 自定义方法
...         'average only odd terms in each column'
...         df = self._df
...         return df[df % 2==1].mean(axis=0)
...
```

现在，我们可以使用以 custom 为名称前缀的新方法，如下所示：

```
>>> df.custom.odds  # 作为属性
   A   C
a  0   2
b  4   6
c  8  10
>>> df.custom.avg_odds()  # 作为方法
A    nan
B    5.00
C    nan
D    7.00
dtype: float64
```

重要的是，除了 custom 之外，你还可以在装饰器中指定任何你想要使用的单词。类似地，register_series_accessor 用于 Series 对象，register_index_accessor 用于 Index 对象[一]，它们都可以实现相同的功能。

5.15 滚动和填充操作

由于 Pandas 在量化金融领域的传统，许多滚动时间序列计算都很容易实现。让我们加载一些股价数据。

```
>>> from pandas_datareader import data
>>> df=data.DataReader("F", 'yahoo', '20200101',
↪     '20200130').reset_index()
>>> df.head()
        Date  High   Low  Open  Close      Volume  Adj Close
0 2020-01-02  9.42  9.19  9.29   9.42 43425700.00       9.26
1 2020-01-03  9.37  9.15  9.31   9.21 45040800.00       9.06
2 2020-01-06  9.17  9.06  9.10   9.16 43372300.00       9.01
3 2020-01-07  9.25  9.12  9.20   9.25 44984100.00       9.10
4 2020-01-08  9.30  9.17  9.23   9.25 45994900.00       9.10
```

我们可以计算后面三个元素的平均值。

```
>>> df.rolling(3).mean().head(5)
   High   Low  Open  Close      Volume  Adj Close
0   nan   nan   nan   nan         nan        nan
1   nan   nan   nan   nan         nan        nan
2  9.32  9.13  9.23  9.26 43946266.67       9.11
3  9.26  9.11  9.20  9.21 44465733.33       9.05
4  9.24  9.12  9.18  9.22 44783766.67       9.07
```

注意，只有在完全填充的滑动窗口中才能得到有效的输出。窗口的端点是否用于计算取决于 closed 关键字参数。除了默认的矩形窗口外，还可以使用其他窗口，如 Blackman 和 Hamming 窗口。df.rolling()函数会产生一个 Rolling 对象，

[一] 第三方数据清理模块 pyjanitor 广泛使用了这种方法。

其中包含诸如 apply、aggregate 等方法。类似于滚动窗口计算，指数加权窗口计算可以使用 ewm（ ）方法进行计算。

```
>>> df.ewm(3).mean()
    High  Low  Open  Close      Volume  Adj Close
0   9.42 9.19  9.29   9.42 43425700.00       9.26
1   9.39 9.17  9.30   9.30 44348614.29       9.14
2   9.30 9.12  9.21   9.24 43926424.32       9.08
3   9.28 9.12  9.21   9.24 44313231.43       9.09
4   9.29 9.14  9.22   9.25 44864456.98       9.09
5   9.29 9.15  9.24   9.25 46979043.75       9.10
6   9.31 9.18  9.25   9.25 44906738.44       9.10
7   9.30 9.16  9.25   9.25 45919910.43       9.09
8   9.31 9.17  9.24   9.26 45113266.20       9.10
9   9.30 9.18  9.25   9.24 47977202.99       9.09
10  9.30 9.17  9.24   9.22 47020077.94       9.07
11  9.28 9.16  9.23   9.21 45632324.49       9.05
12  9.27 9.14  9.21   9.21 46637216.86       9.05
13  9.26 9.15  9.21   9.20 44926124.49       9.04
14  9.24 9.09  9.19   9.18 52761475.78       9.03
15  9.21 9.06  9.17   9.14 56635156.17       8.98
16  9.14 8.99  9.10   9.07 57676520.00       8.92
17  9.11 8.96  9.06   9.05 64587200.41       8.90
18  9.07 8.93  9.01   9.00 63198880.10       8.89
19  9.01 8.88  8.96   8.96 58089908.44       8.88
```

以上的学习仅仅触及了 Pandas 功能的表面，完全忽略了它强大的日期和时间管理功能。还有许多内容需要学习，Pandas 官方网站上的在线文档和教程非常适合进一步学习。

第 6 章

可视化数据

本章将讨论使用关键的 Python 模块来可视化数据，但在我们深入讨论之前，理解数据可视化的具体原则是非常重要的，因为这些模块可以处理绘图的方式，但不能确定绘制的内容。

在构建数据可视化之前，最重要的是要站在用户的角度考虑。这部分很容易出错，因为你（作为作者）通常比用户有更丰富的视觉词汇，再加上你对数据的熟悉，容易忽视用户可能会困惑的许多点。

更糟糕的是，人类感知主要是想象的产物，依赖于弥补我们物理视觉硬件的局限性。这意味着不仅美是用户眼中的事物，你的数据可视化也是如此。优秀的数据科学力求传达客观的事实，但我们对视觉的情感反应是潜意识的，也就是说，在我们有意识地察觉之前就已经对呈现做出了反应。用户通常在事后为他们的最初情感印象寻找理由，而不是通过理性思考得出结论。因此，数据可视化的作者必须时刻警惕这些问题。

本章讨论的内容并不新颖，定量信息的视觉呈现是一个历史悠久的领域，值得单独研究。至少，你应该熟悉数据图形的感知属性，以实现沟通精度的层次结构。在顶部（即最准确的地方）是使用位置来表示数据（即散点图），因为我们人类的视觉认知非常擅长在远处找出点簇。因此，尽可能使用位置来传达视觉中最重要的方面。接下来是长度（即条形图），因为只要图形中的其他所有内容都对齐，我们就可以轻松区分长度差异。长度之后是角度，这解释了为什么飞机驾驶舱的仪表都指向同一方向，它便于飞行员可以从指针的角度偏离中立即发现仪表的异常。接下来是面积，然后是体积，最后是密度。我们很难判断两个圆是否具有相同的面积，除非其中一个明显比另一个大。体积是最困难的，因为它受形状的影响很大。如果你曾经尝试将一壶水倒入一个方形盘子中，那么你就知道这是多么难以判断。颜色也很困难，因为颜色带有强烈的情感负担，可能会分散注意力，而且还有色盲等问题。

因此，关键是找出数据可视化的主要信息，并尽可能使用图形层次结构来编码该信息，首选使用位置。次要信息则可以被归类为层次结构中的角度或颜色等更低级别的项目。然后，从可视化中删除所有不利于传达的信息的内容，包括默认颜色、线条、坐标轴、文本或任何妨碍信息传达的内容。请记住，Python 可视化模块具有其自己的默认设置，这可能不利于你期望信息的传达。

6.1 Matplotlib

Matplotlib 是 Python 中科学图形的主要可视化工具。和所有优秀的开源项目一样，它起源于满足个人需求。在其诞生之时，John Hunter 主要使用 Matlab 进行科学可视化，但随着他开始使用 Python 整合来自不同来源的数据，他意识到需要一个 Python 解决方案来进行可视化，因此编写了 Matplotlib 的初始版本。自那些早期年份的版本开始，Matplotlib 逐渐取代了其他竞争的二维科学可视化方法，如今仍然是一个非常活跃的项目。尽管 John Hunter 于 2012 年不幸去世。此外，像 seaborn 这样的其他项目也利用 Matplotlib 的基本元素进行专门的绘图。通过这种方式，Matplotlib 已经渗透到科学 Python 社区的可视化基础设施中。Matplotlib 的关键和持久优势在于其完整性——几乎你能想到的任何二维科学图形都可以使用 Matplotlib 生成，正如 John Hunter 绘图卓越大赛所展示的那样。

Matplotlib 有两个主要概念组成部分：画布（canvas）和艺术家（artists）。画布可以被视为可视化的目标，而艺术家在画布上绘制。要在 Matplotlib 中创建一个画布，你可以使用下面显示的 plt.figure 函数。然后，plt.plot 函数将 Line2D 艺术家放置在画布上，绘制图 6.1，并作为 Python 列表返回输出。

```
>>> import numpy as np
>>> import matplotlib.pyplot as plt
>>> plt.figure() # 创建画布
<Figure size 640x480 with 0 Axes>
>>> x = np.arange(10) # 创建一些数据
>>> y = x**2
>>> plt.plot(x,y)
[<matplotlib.lines.Line2D object at 0x7f9373184850>]
>>> plt.xlabel('This is the xlabel') # 坐标标签
Text(0.5, 0, 'This is the xlabel')
>>> plt.ylabel('This is the ylabel')
Text(0, 0.5, 'This is the ylabel')
>>> plt.show() # 显示图
```

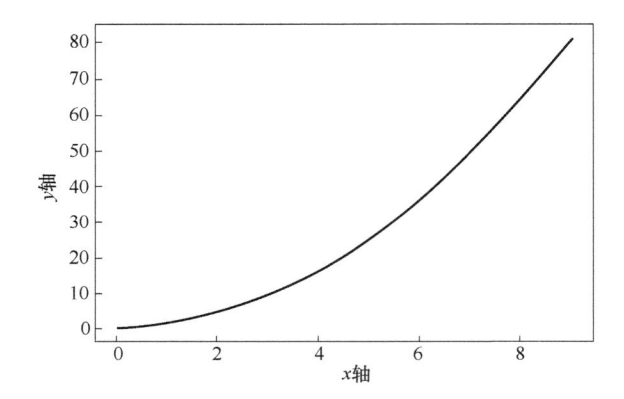

图 6.1 基本 Matplotlib 图

注意，如果在普通 Python 解释器中执行上述操作，则进程将在最后一行冻结（或阻塞）。这是因为 Matplotlib 专注于渲染 GUI 窗口，这是生成图形的最终目标。plt.show 函数触发艺术家在画布上渲染。

虽然使用这些函数是使用 Matplotlib 的传统方式，但面向对象的接口会更有组织性和紧凑性。下面将使用面向对象的风格重新进行上述操作。

```
>>> from matplotlib.pylab import subplots
>>> fig,ax = subplots()
>>> ax.plot(x,y)
>>> ax.set_xlabel('this is xlabel')
>>> ax.set_ylabel('this is ylabel')
```

注意，上面代码的关键步骤是使用 subplots 函数生成图形窗口和坐标轴的对象。然后，绘图命令被附加到相应的 ax 对象上。这使得跟踪叠加在同一个 ax 上的多个可视化变得更容易。

6.1.1 设置默认值

你可以使用 rcParams 字典设置绘图的默认值。

```
>>> import matplotlib
>>> matplotlib.rcParams['lines.linewidth']=2.0 # 线宽设置
```

或者使用关键字参数逐行设置线宽：

```
plot(arange(10),linewidth=2.0) # 使用线宽关键字
```

6.1.2 图例

图例标识绘图上的线条（见图 6.2），loc 表示图例的位置。

```
>>> fig,ax=subplots()
>>> ax.plot(x,y,x,2*y,'or--')
>>> ax.legend(('one','two'),loc='best')
```

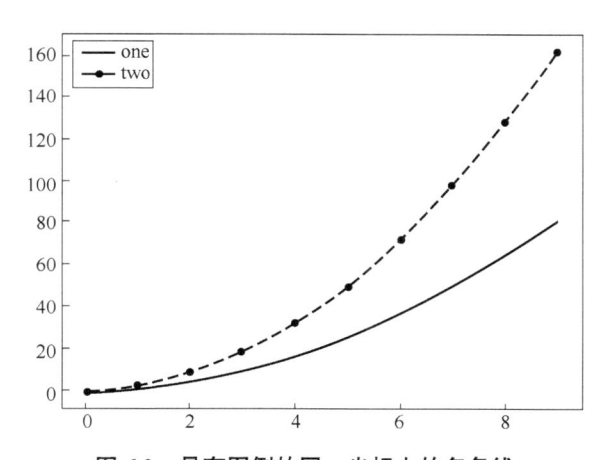

图 6.2 具有图例的同一坐标上的多条线

6.1.3 子图

subplots 函数允许创建多个子图（见图 6.3），每个坐标轴都像一个 Numpy 数组一样被索引。

```
>>> fig,axs = subplots(2,1) # 2 行,1 列
>>> axs[0].plot(x,y,'r-o')
>>> axs[1].plot(x,y*3,'g--s')
```

图 6.3　同一图中的多个子图

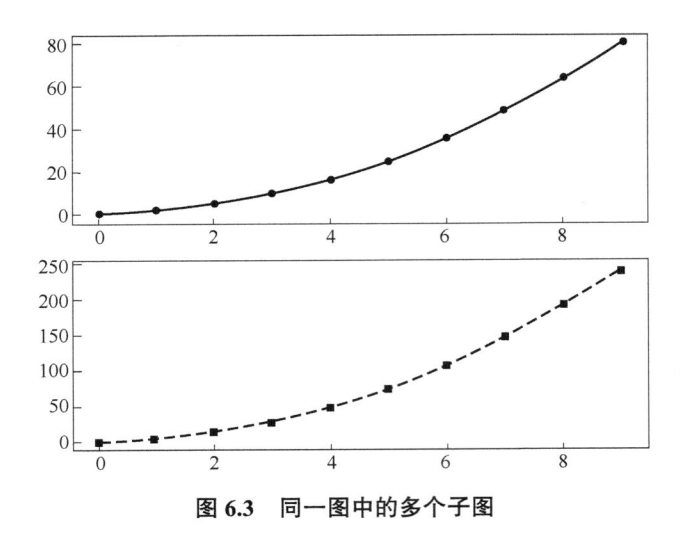

6.1.4　Spines

包含绘图的矩形有四个所谓的轴脊，这些轴脊由 spines 对象管理。在下面的示例中，注意 axs 包含一个可切片的单独坐标轴对象数组（见图 6.4）。左上角的子图对应于 axs[0,0]。对于这个子图，通过将其颜色设置为 'none'，可以使右侧的轴脊不可见。注意，对于 Matplotlib，字符串 'none' 与通常的 Python None 处理方式不同。底部的轴脊通过 set_position 移动到中心，刻度的位置则通过 set_ticks_position 分配。

```
>>> fig,axs = subplots(2,2)
>>> x = np.linspace(-np.pi,np.pi,100)
>>> y = 2*np.sin(x)
>>> ax = axs[0,0]
>>> ax.set_title('centered spines')
>>> ax.plot(x,y)
>>> ax.spines['left'].set_position('center')
>>> ax.spines['right'].set_color('none')
>>> ax.spines['bottom'].set_position('center')
>>> ax.spines['top'].set_color('none')
>>> ax.xaxis.set_ticks_position('bottom')
>>> ax.yaxis.set_ticks_position('left')
```

左下角的子图中，axes[1,0] 将底部轴脊移动到数据的 'zero' 位置。

```
>>> ax = axs[1,0]
>>> ax.set_title('zeroed spines')
>>> ax.plot(x,y)
>>> ax.spines['left'].set_position('zero')
>>> ax.spines['right'].set_color('none')
```

```
>>> ax.spines['bottom'].set_position('zero')
>>> ax.spines['top'].set_color('none')
>>> ax.xaxis.set_ticks_position('bottom')
>>> ax.yaxis.set_ticks_position('left')
```

右上角的子图中，axes[0,1]将底部轴脊移动到轴坐标系的0.1位置，左侧轴脊位于轴坐标系的0.6位置（稍后将详细介绍坐标系）。

```
>>> ax = axs[0,1]
>>> ax.set_title('spines at axes (0.6, 0.1)')
>>> ax.plot(x,y)
>>> ax.spines['left'].set_position(('axes',0.6))
>>> ax.spines['right'].set_color('none')
>>> ax.spines['bottom'].set_position(('axes',0.1))
>>> ax.spines['top'].set_color('none')
>>> ax.xaxis.set_ticks_position('bottom')
>>> ax.yaxis.set_ticks_position('left')
```

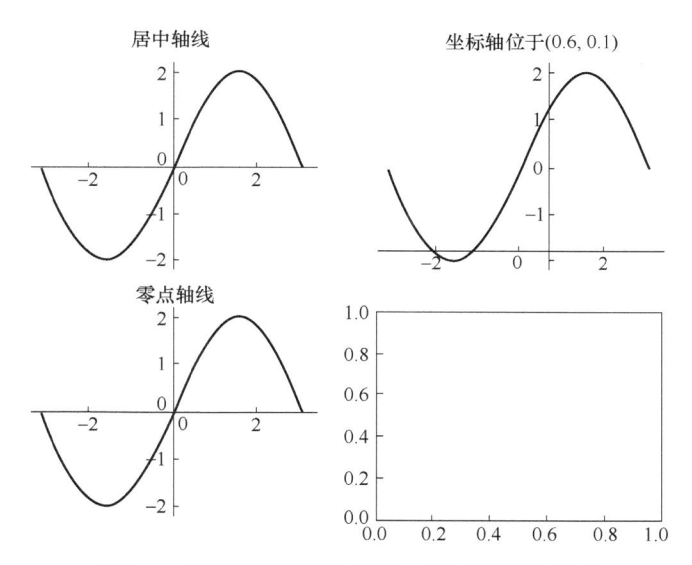

图 6.4　轴脊指的是框架边缘，可以在图中移动

6.1.5　共享轴

subplots函数还允许使用sharex和sharey在不同图之间共享单独的坐标轴（见图6.5），这对于对齐时间序列图特别有帮助。

```
>>> fig, axs = subplots(3,1,sharex=True,sharey=True)
>>> t = np.arange(0.0, 2.0, 0.01)
>>> s1 = np.sin(2*np.pi*t)
>>> s2 = np.exp(-t)
>>> s3 = s1*s2
```

```
>>> axs[0].plot(t,s1)
>>> axs[1].plot(t,s2)
>>> axs[2].plot(t,s3)
>>> ax.set_xlabel('x-coordinate')
```

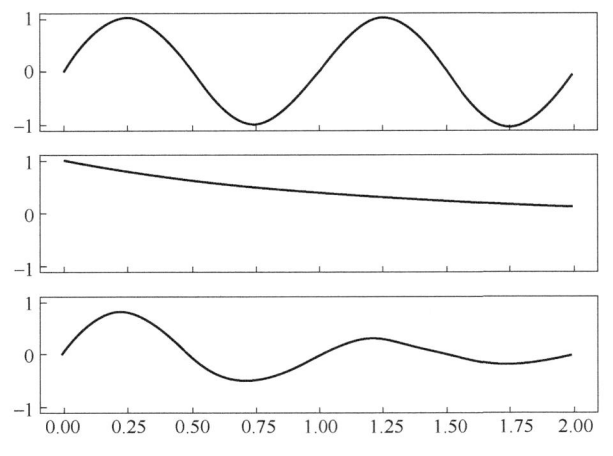

图 6.5 多个子图可以共享一个 *x* 轴

6.1.6 三维曲面

Matplotlib 主要是一个二维绘图软件包，但它具有一些主要基于投影机制的有限三维功能。这方面的主要机制在 mplot3d 子模块中。以下代码绘制了图 6.6。Axes3D 对象具有用于绘制三维图形的 plot_surface 方法，而 cm 模块具有色彩表。

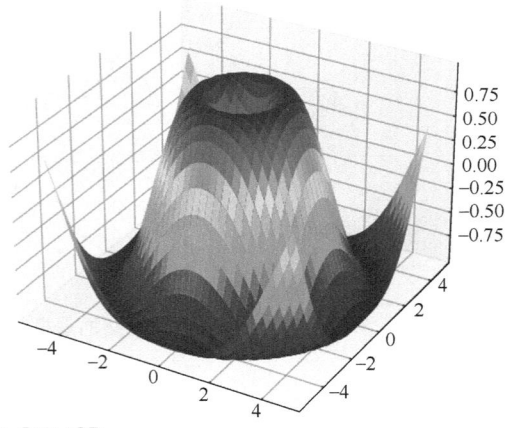

```
>>> from mpl_toolkits.mplot3d import Axes3D
>>> from matplotlib import cm
>>> fig = plt.figure()
>>> ax = Axes3D(fig)
>>> X = np.arange(-5, 5, 0.25)
>>> Y = np.arange(-5, 5, 0.25)
>>> X, Y = np.meshgrid(X, Y)
>>> R = np.sqrt(X**2 + Y**2)
>>> Z = np.sin(R)
>>> ax.plot_surface(X, Y, Z, rstride=1, cstride=1, cmap=cm.jet)
<mpl_toolkits.mplot3d.art3d.Poly3DCollection object at
0x7f9372d6b5b0>
```

图 6.6 三维曲面图

6.1.7 使用 patch

matplotlib.patches 模块中可用于绘制圆、多边形等（见图 6.7）。

```
>>> from matplotlib.patches import Circle, Rectangle, Ellipse
>>> fig,ax = subplots()
>>> ax.add_patch(Circle((0,0), 1,color='g'))
>>> ax.add_patch(Rectangle((-1,-1),
...         width = 2, height = 2,
...         color='r',
...         alpha=0.5))  # 透明度
>>> ax.add_patch(Ellipse((0,0),
...         width = 2, height = 4, angle = 45,
...         color='b',
...         alpha=0.5))   # 透明度
>>> ax.axis('equal')
(-1.795861705, 1.795861705, -1.795861705, 1.795861705)
>>> ax.grid(True)
```

也可以使用交叉图案填充颜色（见图 6.8）。

```
>>> fig,ax = subplots()
>>> ax.add_patch(Circle((0,0),
...              radius=1,
...              facecolor='w',
...              hatch='x'))
>>> ax.grid(True)
>>> ax.set_title('Using cross-hatches',fontsize=18)
```

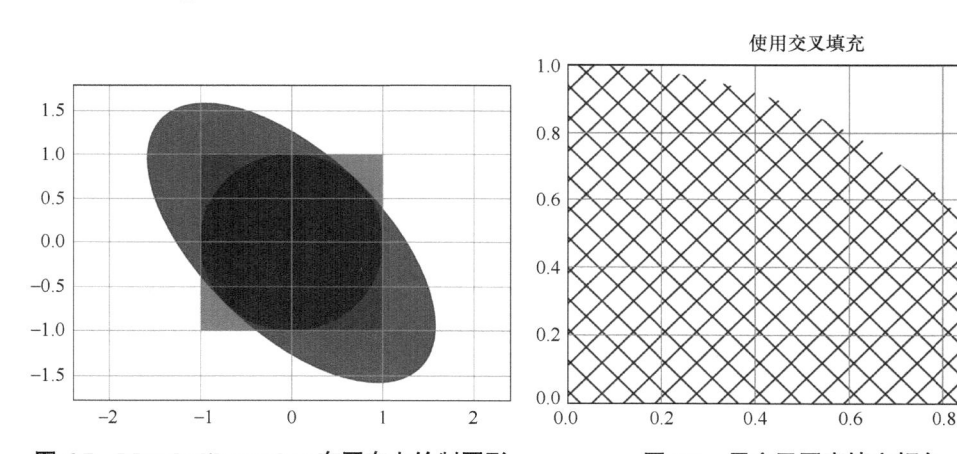

图 6.7　Matplotlib patches 在画布上绘制图形　　　图 6.8　用交叉图案填充颜色

6.1.8 3d 中的 patches

你还可以向三维坐标轴添加 patches。通过向 subplot_kw 关键字参数传递

{'projection':'3d'}字典来创建三维坐标轴。圆形 patches 是按照通常的方式创建的，然后添加到这个坐标轴上。art3d 模块中的 pathpatch_2d_to_3d 方法会改变观察者沿着指定方向对每个 patches 的视角，从而产生了三维效果（见图 6.9）。

```
>>> import mpl_toolkits.mplot3d.art3d as art3d
>>> fig, ax  = subplots(subplot_kw={'projection':'3d'})
>>> c = Circle((0,0),radius=3,color='r')
>>> d = Circle((0,0),radius=3,color='b')
>>> ax.add_patch(c)
>>> ax.add_patch(d)
>>> art3d.pathpatch_2d_to_3d(c,z=0,zdir='y')
>>> art3d.pathpatch_2d_to_3d(d,z=0,zdir='z')
```

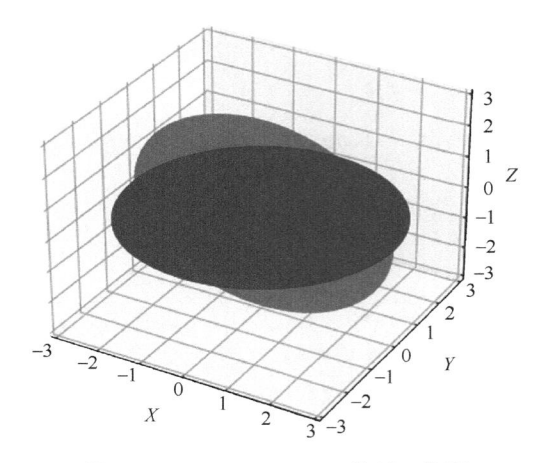

图 6.9 Matplotlib patch 绘制三维图

二维图形也可以在三维中堆叠（见图 6.10）。通过使用 get_paths 提取图形的各个封闭多边形的路径，然后使用这些路径创建 PathPatch 对象并添加到坐标轴中。与之前一样，最后一步是使用 pathpatch_2d_to_3d 在每个 patch 上设置透视视图。

```
>>> from numpy import pi, arange, linspace, sin, cos
>>> from matplotlib.patches import PathPatch
>>> x = linspace(0,2*pi,100)
>>> #从图形创建多边形
>>> fig, ax = subplots()
>>> p1=ax.fill_between(x,sin(x),-1)
>>> p2=ax.fill_between(x,sin(x-pi/3),-1)
>>> path1=p1.get_paths()[0] # 为 p1 获取闭合多边形
>>> path2=p2.get_paths()[0] # 为 p2 获取闭合多边形
>>> ax.set_title('setting up patches from 2D graph')
>>> fig, ax  = subplots(subplot_kw={'projection':'3d'})
>>> pp1 = PathPatch(path1,alpha=0.5) # 需要稍后分配
>>> pp2 = PathPatch(path2,color='r',alpha=0.5) # 需要稍后分配
>>> # 添加 patches
>>> ax.add_patch(pp1)
>>> ax.add_patch(pp2)
```

```
>>> # 变换patches
>>> art3d.pathpatch_2d_to_3d(pp1,z=0,zdir='y')
>>> art3d.pathpatch_2d_to_3d(pp2,z=1,zdir='y')
```

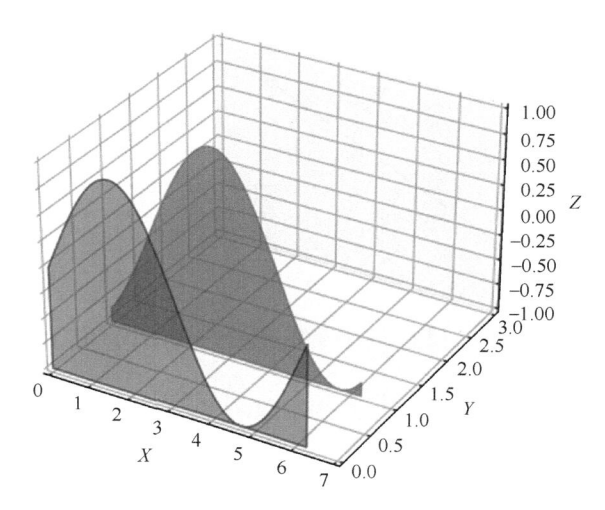

图 6.10 可以从其他 Matplotlib 渲染创建自定义 patch 基元

6.1.9 使用 transformation

Matplotlib 的优势之一是它在创建数据可视化时可以管理多个坐标系。这些坐标系提供了数据和渲染可视化空间之间的比例映射。数据坐标系是数据点的坐标系，而显示坐标系是所显示图形的坐标系。ax.transData 方法将数据坐标 (5, 5) 转换为显示坐标系中的坐标（见图 6.11）。

```
>>> fig,ax = subplots()
>>> # 数据坐标中的线
>>> ax.plot(np.arange(10), np.arange(10))
>>> # 在数据坐标中标记中心
>>> ax.plot(5,5,'o',markersize=10,color='r')
>>> # 在显示坐标系中显示同一点
>>> print(ax.transData.transform((5,5)))
[2560.  1900.8]
```

如果我们使用 figsize 关键字参数创建大小不同的坐标轴，那么即使它们都是指向数据坐标中的同一点，transData 方法也会返回不同的坐标。

```
>>> fig,ax = subplots(figsize=(10,4))
>>> ax.transData.transform((5,5))
array([4000., 1584.])
>>> fig,ax = subplots(figsize=(5,4))
>>> ax.transData.transform((5,5))
array([2000., 1584.])
```

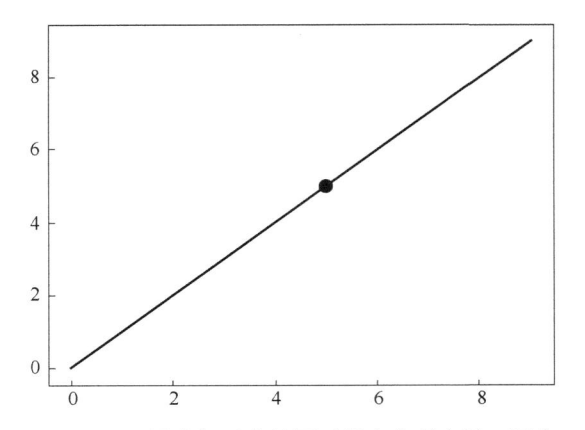

图 6.11 不同坐标系中的图形元素出现在同一图上

此外，如果在 GUI 窗口（而不是 Jupyter Notebook）中绘制此图，并使用鼠标调整图形大小，然后再次执行此操作，那么你也会根据窗口的大小获得不同的坐标。实际上，我们很少在显示坐标中工作。可以使用 ax.transData.inverted ().transform 返回数据坐标。

坐标轴坐标系是包含坐标轴的单位框。transAxes 方法将映射到此坐标系。例如，在图 6.12 中，transAxes 方法作为 transform 关键字参数。

```
>>> from matplotlib.pylab import gca
>>> fig,ax = subplots()
>>> ax.text(0.5,0.5,
...     'Middle of plot',
...     transform = ax.transAxes,
...     fontsize=18)
>>> ax.text(0.1,0.1,
...     'lower left',
...     transform = ax.transAxes,
...     fontsize=18)
>>> ax.text(0.8,0.8,
...     'upper right',
...     transform = ax.transAxes,
...     fontsize=18)
>>> ax.text(0.1,0.8,
...     'upper left',
...     transform = ax.transAxes,
...     fontsize=18)
>>> ax.text(0.8,0.1,
...     'lower right',
...     transform = ax.transAxes,
...     fontsize=18)
```

这对于在绘图中注释时保持一致性非常有用，与数据无关。你可以将此与 patches 结合使用，在图中创建数据坐标和坐标轴坐标混合的项目，如图 6.13 所示，代码如下：

```
>>> fig,ax = subplots()
>>> x = linspace(0,2*pi,100)
>>> ax.plot(x,sin(x) )
>>> ax.add_patch(Rectangle((0.1,0.5),
...        width = 0.5,
...        height = 0.2,
...        color='r',
...        alpha=0.3,
...        transform = ax.transAxes))
>>> ax.axis('equal')
(-0.3141592653589793, 6.5973445725385655,
-1.0998615404412626, 1.0998615404412626)
```

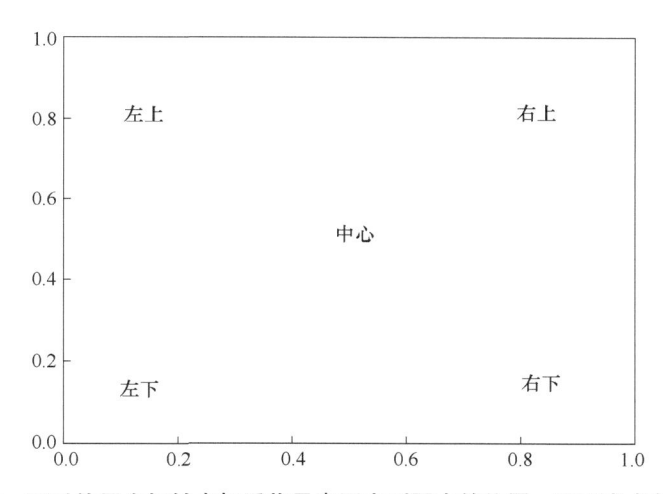

图 6.12 可以使用坐标轴坐标系将元素固定到图中的位置，而不考虑数据坐标

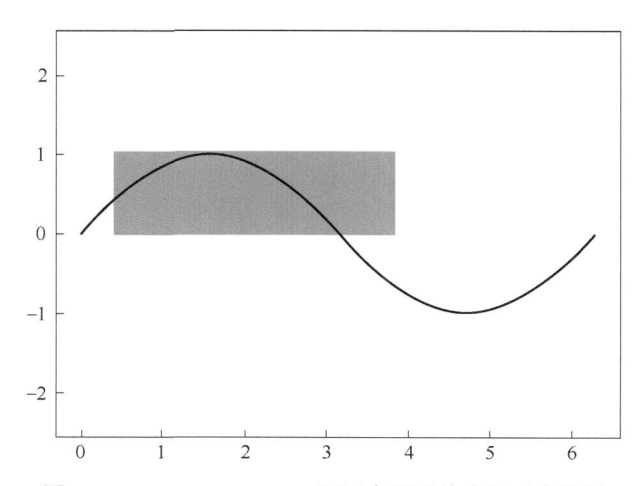

图 6.13 Matplotlib patch 可以在不同的坐标系中使用

6.1.10 使用文本注释

Matplotlib 拥有一个 text 类，该类在可视化中对文本进行操作。使用 ax.text 可以直接添加文本（见图 6.14）。

```
>>> fig,ax = subplots()
>>> x = linspace(0,2*pi,100)
>>> ax.plot(x,sin(x) )
>>> ax.text(pi/2,1,'max',fontsize=18)
>>> ax.text(3*pi/2,-1.1,'min',fontsize=18)
>>> ax.text(pi,0,'zero',fontsize=18)
>>> ax.axis((0,2*pi,-1.25,1.25))
(0.0, 6.283185307179586, -1.25, 1.25)
```

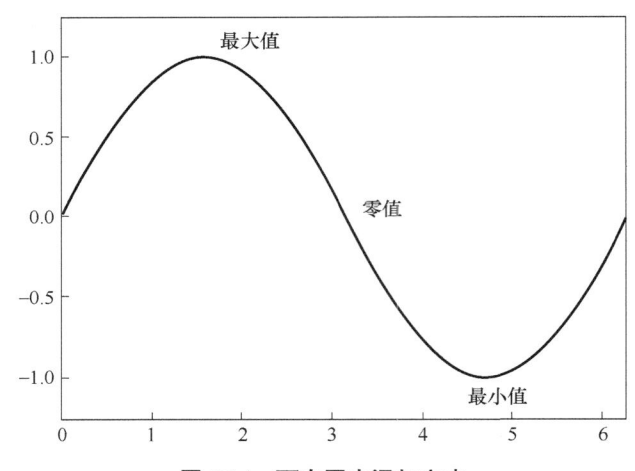

图 6.14 可在图中添加文本

我们还可以使用边界框来包围文本，如图 6.15 所示，这需要使用到 bbox 关键字参数。

```
>>> fig,ax = subplots()
>>> x = linspace(0,2*pi,100)
>>> ax.plot(x,sin(x))
>>> ax.text(pi/2,1-0.5,'max',
...         fontsize=18,
...         bbox = {'boxstyle':'square','facecolor':'yellow'})
```

6.1.11 使用箭头注释

下面的代码中，ax.annotate 实现了用箭头对图进行注释，如下所示（见图 6.16）：

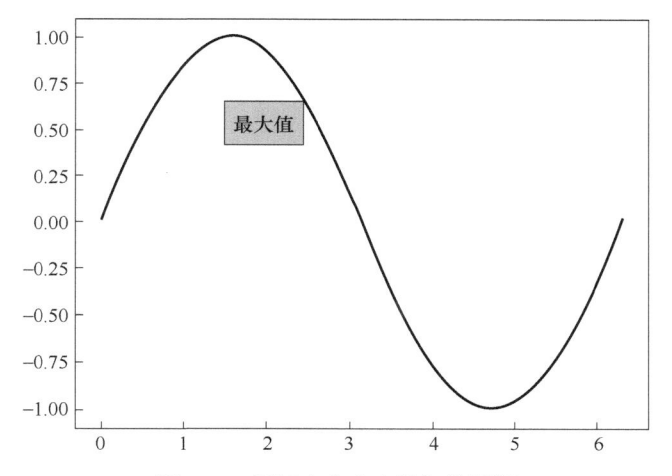

图 6.15 可以在文本上添加边界框

```
>>> fig,ax = subplots()
>>> x = linspace(0,2*pi,100)
>>> ax.plot(x,sin(x))
>>> ax.annotate('max',
...     xy=(pi/2,1),  # 箭头端点的放置位置
...     xytext=(pi/2,0.3),  # 数据坐标中的文字位置
...     arrowprops={'facecolor':'black','shrink':0.05},
...     fontsize=18,
...     )
>>> ax.annotate('min',
...     xy=(3/2.*pi,-1),  # 箭头端点的放置位置
...     xytext=(3*pi/2.,-0.3),  # 数据坐标中的文字位置
...     arrowprops={'facecolor':'black','shrink':0.05},
...     fontsize=18,
...     )
```

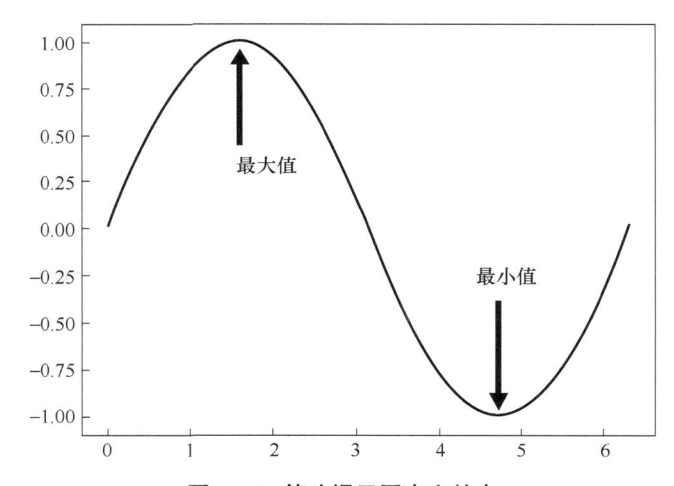

图 6.16 箭头提示图表上的点

我们还可以使用 textcoords 系统明确规定文本坐标，如图 6.17 所示，其中坐标系是用字符串 'axes fraction' 指定的，而不是使用 ax.transAxes。

```
>>> fig,ax = subplots()
>>> x = linspace(0,2*pi,100)
>>> ax.plot(x,sin(x) )
>>> ax.annotate('max',
...     xy=(pi/2,1),        # 箭头端点的放置位置
...     xytext=(0.3,0.8),   # 数据坐标中的文字位置
...     textcoords='axes fraction',
...     arrowprops={'facecolor':'black',
...                 'shrink':0.05},
...     fontsize=18,
...     )
>>> ax.annotate('min',
...     xy=(3/2.*pi,-1),    # 箭头端点的放置位置
...     xytext=(0.8,0.2),   # 数据坐标中的文字位置
...     textcoords='axes fraction',
...     arrowprops={'facecolor':'black',
...                 'shrink':0.05,
...                 'width':10,
...                 'headwidth':20,
...                 'headlength':6},
...     fontsize=18,
...     )
```

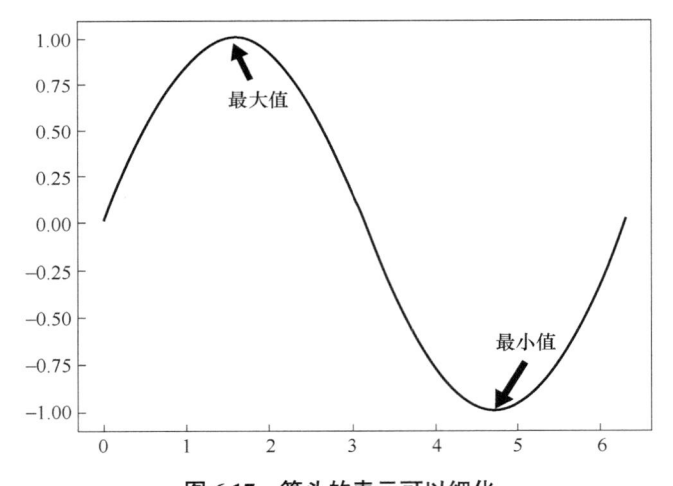

图 6.17 箭头的表示可以细化

有时，我们只需要箭头而不需要文本，如图 6.18 所示，可以用 xytext 给定箭头的尾部坐标，connectionstyle 指定箭头的曲度。

```
>>> fig,ax = subplots()
>>> x = linspace(0,2*pi,100)
>>> ax.set_title('Arrow without text',fontsize=18)
>>> ax.annotate("",  # 将文本字符串参数保留为空
...     xy=(0.2, 0.2), xycoords='data',
...     xytext=(0.8, 0.8), textcoords='data',
```

```
...             arrowprops=dict(arrowstyle="->",
...                             connectionstyle="arc3,rad=0.3",
...                             linewidth=2.0),
...             )
```

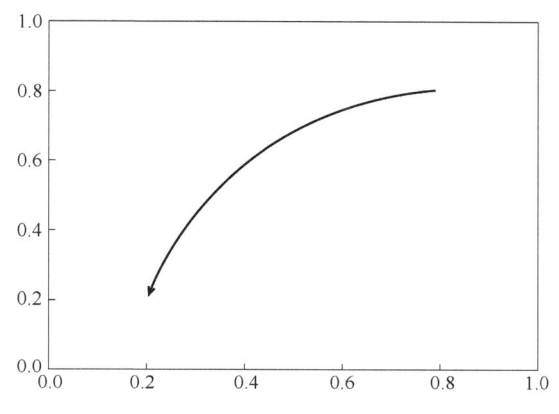

图 6.18 没有文本的箭头

由于箭头对于指向图形元素或表示复杂的矢量场（例如，风速和风向）非常重要，因此 Matplotlib 提供了许多自定义选项。

6.1.12 嵌入可缩放 / 不可缩放的子图

Matplotlib 可实现子图随绘制数据的缩放 / 不缩放。

```
>>> from mpl_toolkits.axes_grid1.anchored_artists import
AnchoredDrawingArea
>>> fig,ax = subplots()
>>> fig.set_size_inches(3,3)
>>> ada = AnchoredDrawingArea(40, 20, 0, 0,
...                           loc=1, pad=0.,
...                           frameon=False)
>>> p1 = Circle((10, 10), 10)
>>> ada.drawing_area.add_artist(p1)
>>> p2 = Circle((30, 10), 5, fc="r")
>>> ada.drawing_area.add_artist(p2)
>>> ax.add_artist(ada)
<mpl_toolkits.axes_grid1.anchored_artists.AnchoredDrawingArea
object at 0x7f939833e760>
```

上面的代码通过 AnchoredDrawingArea 实现在同一图中绘制其他数据而不改变已嵌入的图形，如图 6.19 所示。

下面的代码使用 AnchoredAuxTransformBox 通过数据坐标进行缩放来实现对嵌入图形的缩放。如图 6.20 所示，嵌入的椭圆使用 ax 进行缩放。

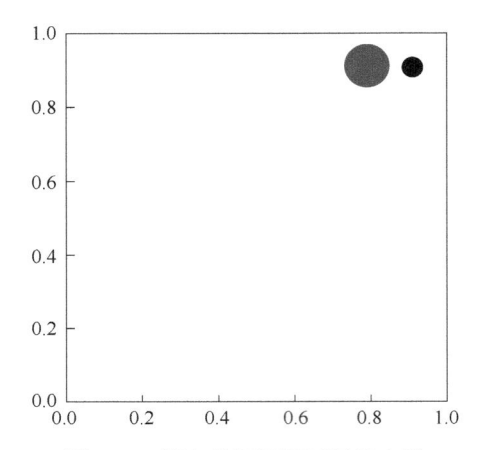

图 6.19　嵌入的子图可以是独立的

```
>>> from matplotlib.patches import Ellipse
>>> from mpl_toolkits.axes_grid1.anchored_artists import
AnchoredAuxTransformBox

>>> fig,ax = subplots()
>>> box = AnchoredAuxTransformBox(ax.transData, loc=2)
>>> # 数据坐标
>>> el = Ellipse((0,0),
...         width=0.1,
...         height=0.4,
...         angle=30)
>>> box.drawing_area.add_artist(el)
>>> ax.add_artist(box)
<mpl_toolkits.axes_grid1.anchored_artists.AnchoredAuxTransformBox
object at 0x7f9372f5dbe0>
```

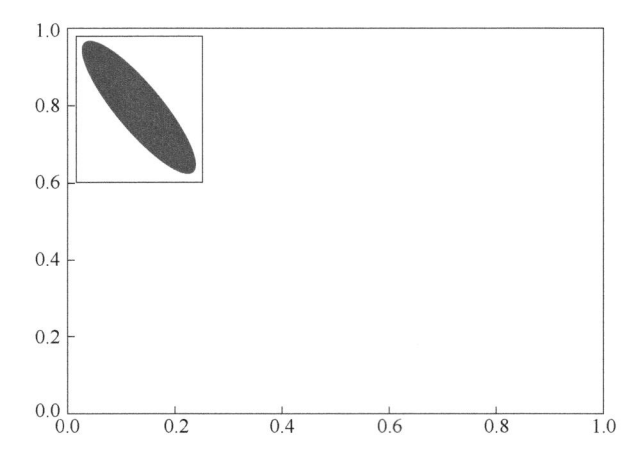

图 6.20　与图 6.19 不同的是，这个子图会随着图中其他数据图形的缩放而改变

6.1.13 动画

创建动画的一种方法是在迭代中生成艺术家序列,然后对其进行动画处理(见图 6.21)。

```
>>> import matplotlib.animation as animation
>>> fig,ax = subplots()
>>> x = np.arange(10)
>>> frames = [ax.plot(x,x,x[i],x[i],'ro',ms=5+i*10)
...           for i in range(10)]
>>> # 必须在下一行指定
>>> g=animation.ArtistAnimation(fig,frames,interval=50)
```

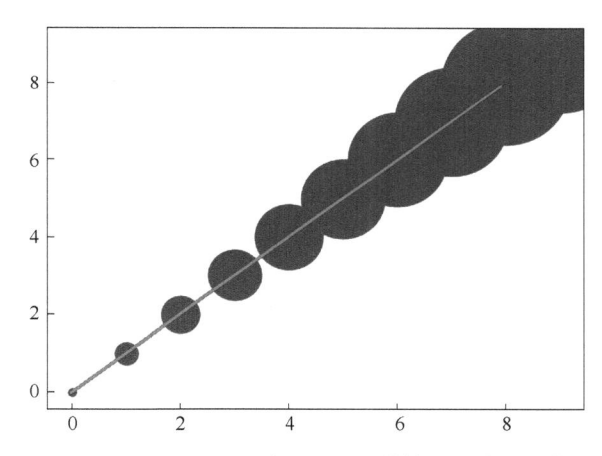

图 6.21　**Matplotlib** 动画利用了其他图形基础元素

上面这种方法适用于相对较少的帧。当然,也可以动态地创建帧(见图 6.22)。

```
>>> import matplotlib.animation as animation
>>> x = np.arange(10)
>>> linewidths =[10,20,30]
>>> fig = plt.figure()
>>> ax = fig.add_subplot(111)
>>> line, = ax.plot(x,x,'-ro',ms=20,linewidth=5.0)
>>> def update(data):
...     line.set_linewidth(data)
...     return (line,)
...
>>> ani = animation.FuncAnimation(fig, update, x, interval=500)
>>> plt.show()
```

要在 Jupyter Notebook 中制作这些动画,必须用 to_jshtml 将它们转换为相应的 JavaScript 动画。例如,在一个单元格中,执行以下操作(见图 6.23):

```
>>> fig,ax = subplots()
>>> frames = [ax.plot(x,x,x[i],x[i],'ro',ms=5+i*10)
```

```
...                     for i in range(10)]
>>> g=animation.ArtistAnimation(fig,frames,interval=50)
from IPython.display import HTML
HTML(g.to_jshtml())
```

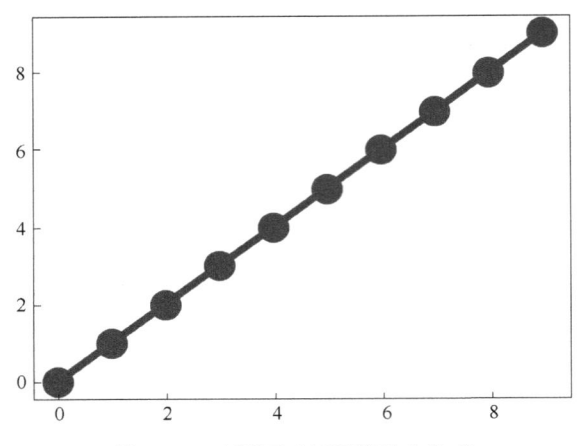

图 6.22 动画可以用函数动态生成

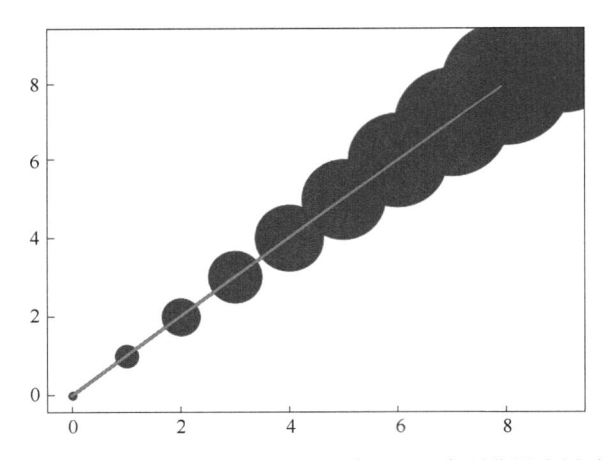

图 6.23 Jupyter Notebook 中的动画可以在浏览器中播放

6.1.14 直接使用路径

通过使用路径，可以在 Matplotlib 中对绘图进行更低级别的访问，你可以将其视为对正在画布上绘图的指示笔进行编程。例如，patch 由路径组成。路径具有顶点和相应的绘制命令。例如，对于图 6.24，这将在两点之间绘制一条线。

```
>>> from matplotlib.path import Path
>>> vertices=[ (0,0),(1,1) ]
>>> codes = [ Path.MOVETO, # 将触笔移动到(0,0)
...           Path.LINETO] # 画线至(1,1)
>>> path = Path(vertices, codes)   # 创建路径
>>> fig, ax = subplots()
>>> # 转换路径到patch
>>> patch = PathPatch(path,linewidth=10)
>>> ax.add_patch(patch)
>>> ax.set_xlim(-.5,1.5)
(-0.5, 1.5)
>>> ax.set_ylim(-.5,1.5)
(-0.5, 1.5)
>>> plt.show()
```

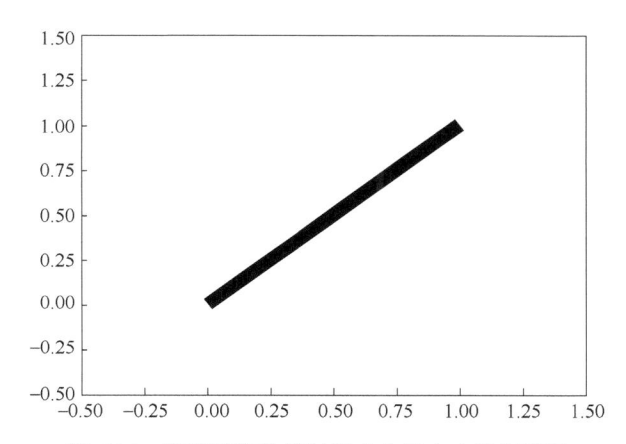

图 6.24 使用特定的笔触指令在画布上绘制路径

路径也可以使用二次和三次贝塞尔曲线来连接点，如图 6.25 所示。

```
>>> vertices=[ (-1,0),(0,1),(1,0),(0,-1),(-1,0) ]
>>> codes = [Path.MOVETO, # 将触笔移动到(0,0)
...          Path.CURVE3, # 绘制曲线
...          Path.CURVE3, # 绘制曲线
...          Path.CURVE3, # 绘制曲线
...          Path.CURVE3,]
>>> path = Path(vertices, codes)   # 创建路径
>>> fig, ax = subplots()
>>> # 转换路径到patch
>>> patch = PathPatch(path,linewidth=2)
>>> ax.add_patch(patch)
>>> ax.set_xlim(-2,2)
(-2.0, 2.0)
>>> ax.set_ylim(-2,2)
(-2.0, 2.0)
>>> ax.set_title('Quadratic Bezier Curve Path')
>>> for i in vertices:
...     _=ax.plot(i[0],i[1],'or'  )# 控制点
...     _=ax.text(i[0],i[1],'control\n
↪   point',horizontalalignment='center')
...
```

二次贝塞尔曲线路径

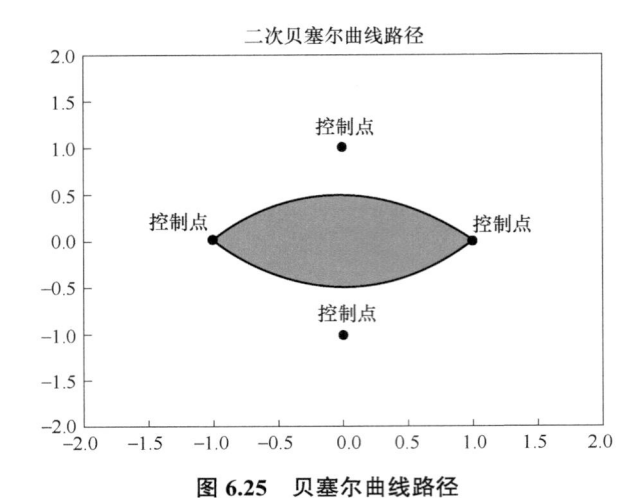

图 6.25 贝塞尔曲线路径

箭头和曲线可以组合以显示曲线上各点的方向导数（见图 6.26）。

```
>>> x = np.linspace(0,2*np.pi,100)
>>> y = np.sin(x)
>>> fig, ax = subplots()
>>> ax.plot(x,y)
>>> u = []
>>> # 子样本 x
>>> x = x[::10]
>>> for i in zip(np.ones(x.shape),np.cos(x)):
...     v=np.array(i)
...     u.append(v/np.sqrt(np.dot(v,v)) )
...
>>> U=np.array(u)
>>> ax.quiver(x,np.sin(x),U[:,0],U[:,1])
>>> ax.grid()
>>> ax.axis('equal')
(-0.3141592653589793, 6.5973445725385655, -1.0998615404412626,
1.0998615404412626)
```

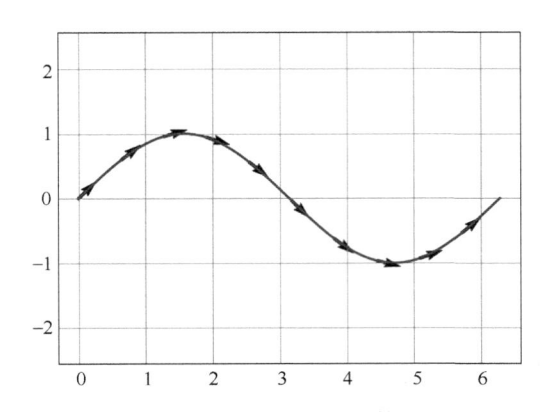

图 6.26 与曲线相切的箭头

6.1.15 使用滑块与绘图交互

Matplotlib GUI 小组件（QT 或其他后端）可以通过回调来响应按键或鼠标移动。这使得在 Matplotlib 中添加基本交互图变得很容易。这些也可以在 Jupyter Notebook 中的 matplotlib.widgets 中获得。使用 on_changed 或 on_clicked 将小组件附加到回调函数 update 上，对特定小组件的更改会触发回调（见图 6.27），其中 Slider、RadioButtons 和 Button 小组件附加到回调函数。

```
>>> from matplotlib.widgets import Slider, Button, RadioButtons
>>> from matplotlib.pylab import subplots_adjust, axes
>>> fig, ax = subplots()
>>> subplots_adjust(left=0.25, bottom=0.25)
>>> # 设置数据
>>> t = arange(0.0, 1.0, 0.001)
>>> a0, f0 = 5, 3
>>> s = a0*sin(2*pi*f0*t)
>>> # 画图
>>> l, = ax.plot(t,s, lw=2, color='red')
>>> ax.axis([0, 1, -10, 10])
(0.0, 1.0, -10.0, 10.0)
>>> axcolor = 'lightgoldenrodyellow'
>>> # 创建 widgets 的轴
>>> axfreq = axes([0.25, 0.1, 0.65, 0.03], facecolor=axcolor)
>>> axamp = axes([0.25, 0.15, 0.65, 0.03], facecolor=axcolor)
>>> sfreq = Slider(axfreq, 'Freq', 0.1, 30.0, valinit=f0)
>>> samp = Slider(axamp, 'Amp', 0.1, 10.0, valinit=a0)
>>> def update(val):
...     amp = samp.val
...     freq = sfreq.val
...     l.set_ydata(amp*sin(2*pi*freq*t))
...     draw()
...
>>> # 将回调附加到 widgets
>>> sfreq.on_changed(update)
0
>>> samp.on_changed(update)
0
>>> resetax = axes([0.8, 0.025, 0.1, 0.04])
>>> button = Button(resetax, 'Reset', color=axcolor,
↪       hovercolor='0.975')
>>> def reset(event):
...     sfreq.reset()
...     samp.reset()
...
>>> # 回调 button
>>> button.on_clicked(reset)
0
>>> rax = axes([0.025, 0.5, 0.15, 0.15], facecolor=axcolor)
>>> radio = RadioButtons(rax, ('red', 'blue', 'green'), active=0)
>>> def colorfunc(label):
...     l.set_color(label)
```

```
...        draw()
...
>>> # 回调radio buttons
>>> radio.on_clicked(colorfunc)
0
```

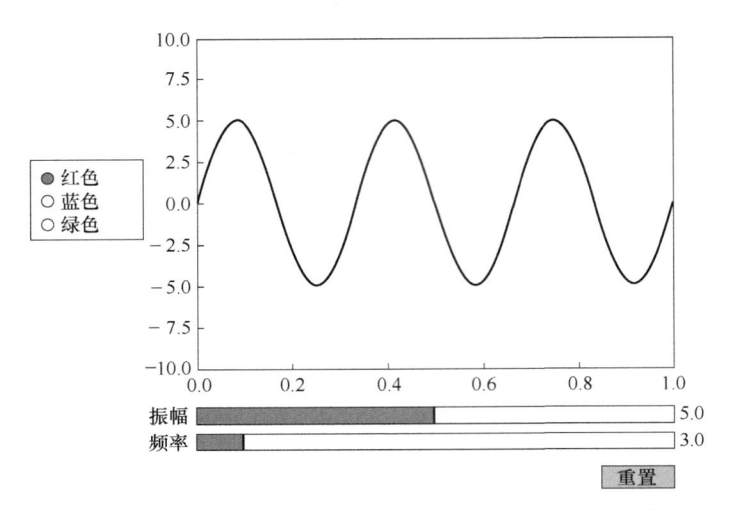

图 6.27 交互式小组件可以附加到 Matplotlib GUI 后端

6.1.16 色彩图

Matplotlib 提供了许多有用的色彩图。imshow 函数接受一个输入数组，并使用与每个条目的值对应的颜色绘制该数组中的单元格（见图 6.28）。

```
>>> fig, ax   = subplots()
>>> x = np.linspace(-1,1,100)
>>> y = np.linspace(-3,1,100)
>>> ax.imshow(abs(x + y[:,None]*1j)) # 使用广播
```

Matplotlib 的颜色被组织在 matplotlib.colors 子模块中，matplotlib.pylab.cm 模块具有色彩图的接口，对于 cm 中的每个色彩，都存在一个反向色彩。例如，cm.Blues 和 cm.Blues_r。我们可以将此色彩图与 imshow 一起使用，使用 cmap 关键字参数创建图 6.29。

```
>>> fig, ax   = subplots()
>>> ax.imshow(abs(x + y[:,None]*1j),cmap=cm.Blues)
```

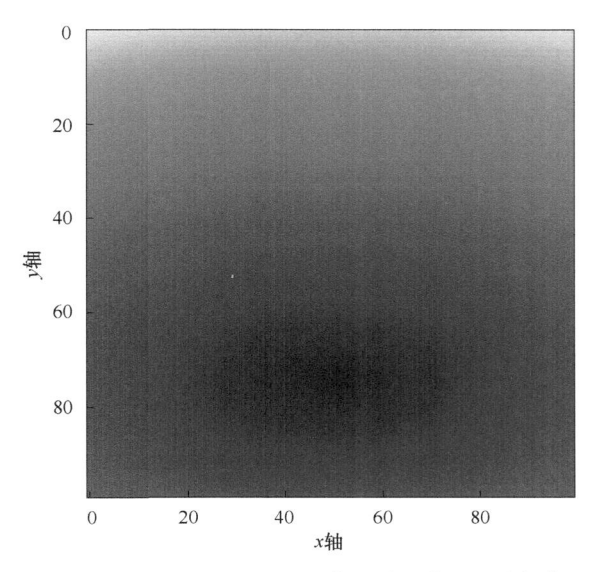

图 6.28　Matplotlib imshow 将矩阵元素显示为颜色

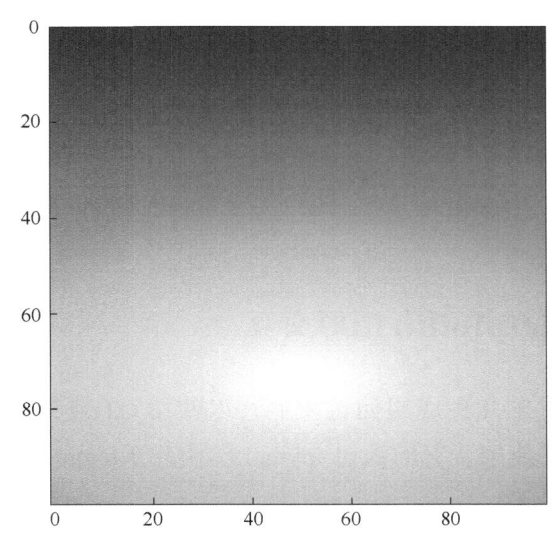

图 6.29　与图 6.28 相同，但使用 cm.Blues 颜色图

6.1.17　使用 setp 和 getp

因为 Matplotlib 元素很多，所以对应的属性也有很多，getp 和 setp 函数可用来更改任何特定元素的属性（见图 6.30）。

```
>>> from matplotlib.pylab import setp, getp
>>> fig, ax = subplots()
>>> c = ax.add_patch(Circle((0,0),1,facecolor='w',hatch='-|'))
>>> # 将轴背景矩形设置为蓝色
>>> setp(ax.get_children()[-1],fc='lightblue')
[None]
>>> ax.set_title('upper right corner is axes background')
>>> ax.set_aspect(1)
>>> fig.show()
```

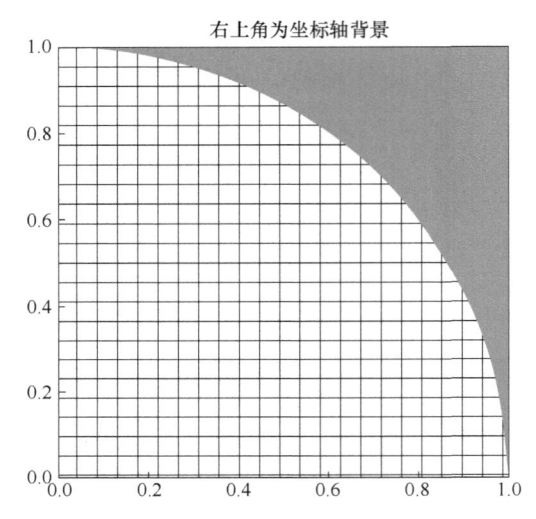

图 6.30 Matplotlib 元素的属性可以使用 **getp** 和 **setp** 发现和更改

6.1.18　与 Matplotlib 图形交互

Matplotlib 提供了几种底层回调机制，以使用 QT 或其他 GUI 后端来扩展图形窗口的交互式使用。它们通过 ipympl 模块和 %matplotlib widget 与 Jupyter Notebook 一起工作。

6.1.19　键盘事件

Matplotlib 的 GUI 图形窗口能够监听并响应键盘事件（即键入的按键）。mpl_connect 和 mpl_disconnect 函数将监听器附加到 GUI 窗口中的事件，并在检测到更改时触发相应的回调。sys.stdout.flush（ ）用来清除标准输出。

```python
import sys
fig, ax = plt.subplots()
# 断开默认连接
fig.canvas.mpl_disconnect(fig.canvas.manager.key_press_handler_id)
ax.set_title('Keystroke events',fontsize=18)
def on_key_press(event):
    print (event.key)
    sys.stdout.flush()

# 将函数连接到画布并监听
fig.canvas.mpl_connect('key_press_event', on_key_press)
plt.show()
```

现在可以在图形窗口中键入按键，并在终端中看到一些输出。我们可以通过引用回调，使图形窗口中的艺术家交互，如下所示：

```python
fig, ax = plt.subplots()
x = np.arange(10)
line, = ax.plot(x, x*x,'-o') # 获取Line2D对象
ax.set_title('More Keystroke events',fontsize=18)
def on_key_press(event):
# 如果event.key是某种简写颜色表示法,则设置线条颜色
    if event.key in 'rgb':
        line.set_color(event.key)
        fig.canvas.draw() # 强制重画

# 断开默认连接
fig.canvas.mpl_disconnect(fig.canvas.manager.key_press_handler_id)
# 将函数连接到画布并监听
fig.canvas.mpl_connect('key_press_event', on_key_press)
plt.show()
```

还可以通过添加另一个回调来增加标记大小，如下所示：

```python
def on_key_press2(event):
# 如果key是某种简写颜色表示法,则设置线条颜色
    if event.key in '123':
        val = int(event.key)*10
        line.set_markersize(val)
        fig.canvas.draw() # 强制重画

fig.canvas.mpl_connect('key_press_event', on_key_press2)
```

还可以使用alt、ctrl、shift等修饰键。

```python
import re
def on_key_press3(event):
    'alt+1,alt+2, changes '
    if re.match('alt\+?',event.key):
        key,=re.match('alt\+(.?)',event.key).groups(0)
        val = int(key)/5.
        line.set_mew(val)
        fig.canvas.draw() # 强制重画

fig.canvas.mpl_connect('key_press_event', on_key_press3)
plt.show()
```

现在，你可以在输入给定的数字、字母和修饰键来更改嵌入的行。

6.1.20 鼠标事件

Matplotlib 的 GUI 图形窗口同样能够监听并响应鼠标事件（移动鼠标、单击鼠标）。这些事件可以用图形 / 数据坐标和按钮的整数 ID（即，左 / 中 / 右键）来响应。

```
fig, ax = plt.subplots()
# 断开默认处理程序
fig.canvas.mpl_disconnect(fig.canvas.manager.key_press_handler_id)

def on_button_press(event):
    button_dict = {1:'left',2:'middle',3:'right'}
    print ("clicked %s button" % button_dict[ event.button ])
    print ("figure coordinates:", event.x, event.y)
    print ("data coordinates:", event.xdata, event.ydata)
    sys.stdout.flush()

fig.canvas.mpl_connect('button_press_event', on_button_press)
plt.show()
```

在每个鼠标单击点上放置点。

```
fig, ax = plt.subplots()
ax.axis([0,1,0,1])
# 断开默认处理程序
fig.canvas.mpl_disconnect(fig.canvas.manager.key_press_handler_id)

o=[]
def on_button_press(event):
    button_dict = {1:'left',2:'middle',3:'right'}
    ax.plot(event.xdata,event.ydata,'o')
    o.append((event.xdata,event.ydata))
    sys.stdout.flush()
    fig.canvas.draw()
fig.canvas.mpl_connect('button_press_event', on_button_press)
plt.show()
```

除了单击鼠标事件之外，还可以将鼠标放在画布上的艺术家上，然后单击它以访问该艺术家。这就是所谓的 pick 事件。

```
fig, ax = plt.subplots()
ax.axis([-1,1,-1,1])
ax.set_aspect(1)
for i in range(5):
    x,y= np.random.rand(2).T
    circle = Circle((x, y),radius=0.1 , picker=True)
    ax.add_patch(circle)

def on_pick(event):
    artist = event.artist
    artist.set_fc(np.random.random(3))
    fig.canvas.draw()

fig.canvas.mpl_connect('pick_event', on_pick)
plt.show()
```

通过单击图中的圆圈，可以随机改变其对应的颜色。注意，在创建图形元素时，必须设置 picker = True 关键字参数。此外，还可以使用 help（plt.connect）查看其他事件的用法。

6.2 Seaborn

Seaborn 是建立在 Matplotlib 之上的，它可以针对 Matplotlib 支持的任何输出进行操作。Seaborn 简化了专门的绘图，使其能够快速、轻松地可视化评估数据的统计方面。我们可以按照通常的方式导入 Seaborn，也可以加载传统的 plt 接口到 Matplotlib。Seaborn 编写非常出色，具有出色的文档和易于阅读的源代码。

```
>>> import pandas as pd
>>> import matplotlib.pyplot as plt
>>> import seaborn as sns
```

Seaborn 提供了许多有趣的数据集。

```
>>> tips = sns.load_dataset('tips')
>>> tips.head()
   total_bill   tip     sex smoker  day    time  size
0       16.99  1.01  Female     No  Sun  Dinner     2
1       10.34  1.66    Male     No  Sun  Dinner     3
2       21.01  3.50    Male     No  Sun  Dinner     3
3       23.68  3.31    Male     No  Sun  Dinner     2
4       24.59  3.61  Female     No  Sun  Dinner     4
>>> sns.relplot(x='total_bill',y='tip',data=tips)
<seaborn.axisgrid.FacetGrid object at 0x7f9370b261c0>
```

我们可以使用 sns.relplot（）轻松地研究变量之间的关系，如图 6.31 所示。

虽然我们可以使用 scatter 在普通的 Matplotlib 中轻松地重新创建此绘图，但 Seaborn 的做法是传入 Pandas 的 Dataframe tips 中，并通过 x 和 y 关键字参数引用要绘制的列。此外，Seaborn 的背景和框架在风格上与 Matplotlib 默认设置不同。重要的是，Seaborn 返回的结果对象是一个 seaborn.axisgrid. FacetGrid 实例，因此如果我们想要检

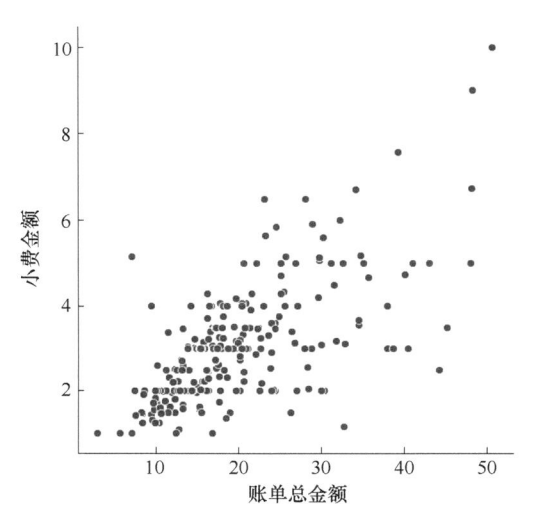

图 6.31 小费数据散点图

索原生的 Matplotlib 元素，则必须从 FacetGrid 中提取它们作为属性，例如 fig 和 ax。

将 smoker 列指定为 hue 角色后，生成的图与之前相同，但是每个圆圈的颜色均由数据框中的 smoker 分类列决定，如图 6.32 所示。

```
>>> sns.relplot(x='total_bill',hue='smoker',y='tip',data=tips)
<seaborn.axisgrid.FacetGrid object at 0x7f9370c3f400>
```

我们也可以将 smoker 列分配给 style 角色来改变标记的形状，如图 6.33 所示。

```
>>> sns.relplot(x='total_bill',y='tip',
...             style='smoker', # 不同的标记形状
...             data=tips)
<seaborn.axisgrid.FacetGrid object at 0x7f9372fe1700>
```

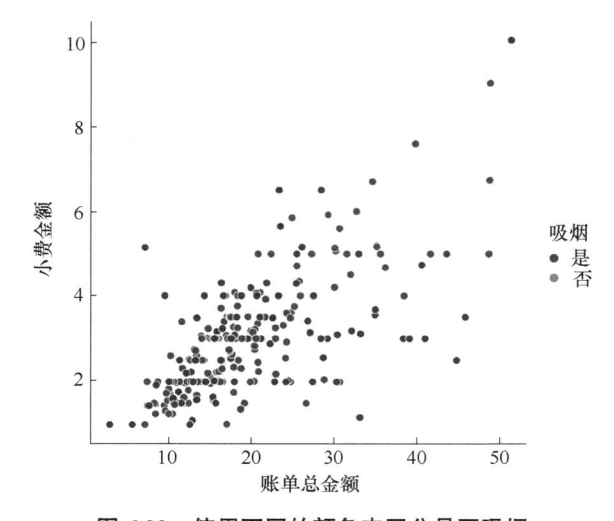

图 6.32　使用不同的颜色来区分是否吸烟

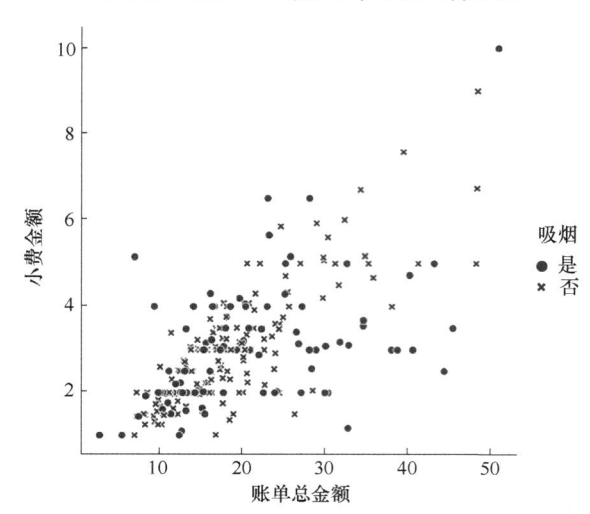

图 6.33　使用不同的标记来区分是否吸烟

你还可以通过指定 markers 关键字参数，为图 6.34 中 smoker 列的两个分类值中的每一个指定有效的 Matplotlib 标记类型。

```
>>> sns.relplot(x='total_bill',y='tip',
...             style='smoker',
...             markers=['s','^'],data=tips)
<seaborn.axisgrid.FacetGrid object at 0x7f9376c6c310>
```

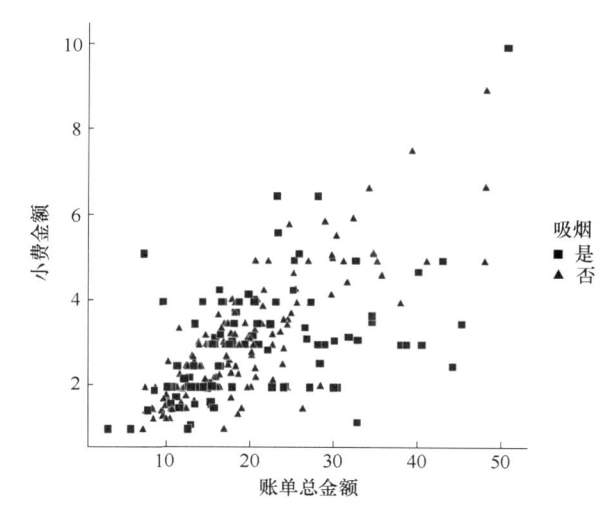

图 6.34　用自定义标记来区分是否吸烟

输入的 size 列可用于缩放标记，如图 6.35 所示。

```
>>> sns.relplot(x='total_bill',y='tip',
...             size='size', # 缩放标记
...             data=tips)
<seaborn.axisgrid.FacetGrid object at 0x7f93713bd310>
```

这是离散的，因为数据框中该列的值是唯一的，

```
>>> tips['size'].unique()
array([2, 3, 4, 1, 6, 5])
```

size 在由关键字参数指定的边界之间进行缩放，例如 sizes=(5,10)。注意，如果为 size 关键字参数提供一个连续的数值列（例如 total_bill）将自动对该列进行离散处理。每个点的透明度值由 alpha 关键字参数指定，该参数传递给 Matplotlib 进行渲染，如图 6.36 所示。

```
>>> sns.relplot(x='total_bill',y='tip',alpha=0.3,data=tips)
<seaborn.axisgrid.FacetGrid object at 0x7f937127b640>
```

此外，由于 Seaborn 处理了 size 关键字参数，所以无法将 scatter size 关键字参数直接传递到最终的 Matplotlib 渲染中。注意，Seaborn 支持对生成的图形外观进行特定的编辑更改，但这些更改很难在 Seaborn 之外进行调整。散点图是

175

relplot（）的默认图类型，但可以通过 kind='line' 参数改为使用折线图。

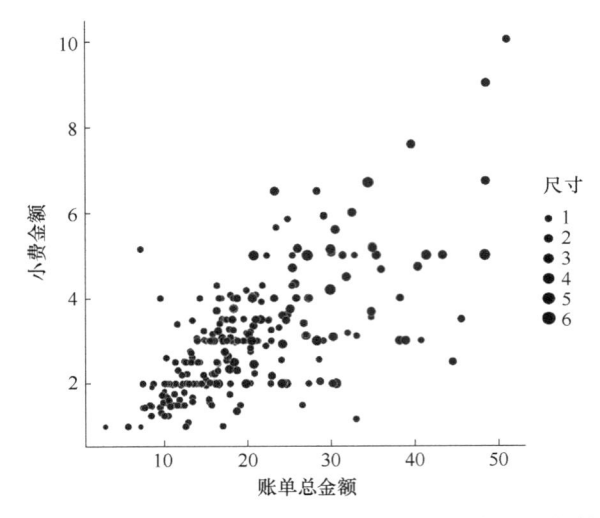

图 6.35 使用 tips（小费）数据集中的 size 列对标记进行缩放

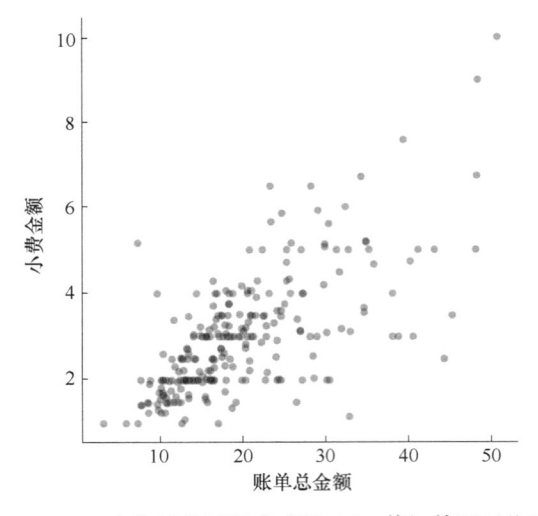

图 6.36 Seaborn 未使用的透明度（即 alpha 值）等可以作为关键字参数传递给 Matplotlib 渲染器

6.2.1 自动聚合

如果一个数据集有多个对应不同 y 值的 x 值，那么 Seaborn 将会聚合这些值。

```
>>> fmri = sns.load_dataset('fmri')
>>> fmri.head()
  subject  timepoint event    region  signal
0     s13         18  stim  parietal   -0.02
1      s5         14  stim  parietal   -0.08
2     s12         18  stim  parietal   -0.08
3     s11         18  stim  parietal   -0.05
4     s10         18  stim  parietal   -0.04
```

每个 timepoint 值对应 56 个 signal 值。

```
>>> fmri['timepoint'].value_counts()
18    56
8     56
1     56
2     56
3     56
4     56
5     56
6     56
7     56
9     56
17    56
10    56
11    56
12    56
13    56
14    56
15    56
16    56
0     56
Name: timepoint, dtype: int64
```

以下是 timepoint=0 组的统计数据：

```
>>> fmri.query('timepoint==0')['signal'].describe()
count    56.00
mean     -0.02
std       0.03
min      -0.06
25%      -0.04
50%      -0.02
75%       0.00
max       0.07
Name: signal, dtype: float64
```

由于这种多重性，Seaborn 将绘制每个点的平均值，用 marker='o' 表示，以及平均估计值的 95% 的置信区间，如图 6.37 所示。

```
>>> sns.relplot(x='timepoint', y='signal', kind='line',data=fmri,
↪     marker='o')
<seaborn.axisgrid.FacetGrid object at 0x7f93709c5df0>
```

这种方法的缺点在于估计值是使用自举法计算的，这对于大型数据集来说可能很慢。因此，可以使用 ci=None 关键字参数关闭自举法，或者用 ci='sd' 替换为标准差计算。设置 estimator=None 关键字参数可以完全停止计算估计值。与 kind='scatter' 类似，可以使用其他数据框列来区分生成的图形。下面使用

hue='event' 为 fmri.event 的每个类别绘制不同的线图：

```
>>> fmri.event.unique()
array(['stim', 'cue'], dtype=object)
```

现在，我们为每个 fmri.event 分类绘制了不同的线图，如图 6.38 所示。

```
>>> sns.relplot(x='timepoint', y='signal',
...             kind='line', data=fmri,
...             hue='event', marker='o')
<seaborn.axisgrid.FacetGrid object at 0x7f93709c5df0>
```

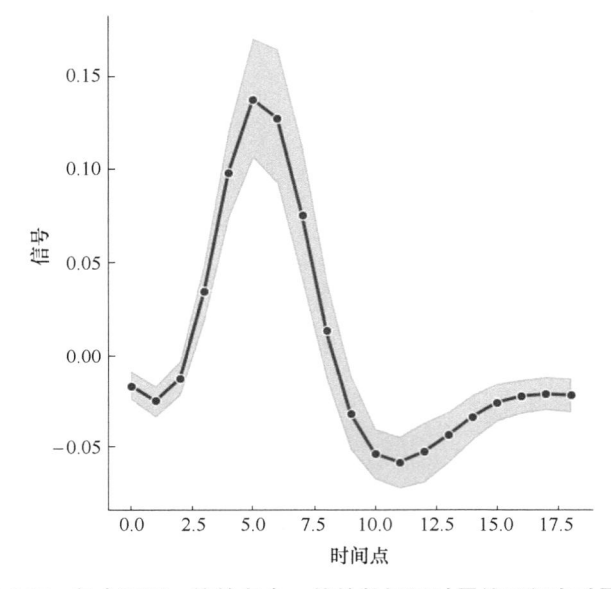

图 6.37　包含不同 y 值的多个 x 值的数据通过置信区间自动聚合

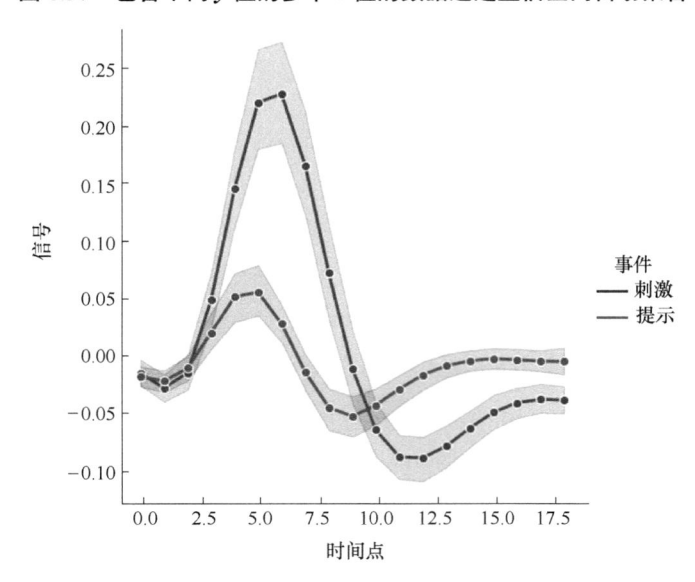

图 6.38　event（事件）数据分类

当 hue 列是数字的而不是分类的,那么颜色将在一个连续的间隔上缩放,如图 6.39 所示。

```
>>> dots = sns.load_dataset('dots').query('align == "dots"')
>>> dots.head()
  align choice  time  coherence  firing_rate
0  dots     T1   -80       0.00        33.19
1  dots     T1   -80       3.20        31.69
2  dots     T1   -80       6.40        34.28
3  dots     T1   -80      12.80        32.63
4  dots     T1   -80      25.60        35.06
>>> sns.relplot(x='time', y='firing_rate',
...             hue='coherence', style='choice',
...             kind='line', data=dots)
<seaborn.axisgrid.FacetGrid object at 0x7f9370883cd0>
```

图 6.39 中有两种不同的线条样式,这是因为 style='choice' 中的 choice 列有两个不同的值。Seaborn 的颜色调色板功能非常强大,以下代码基于 n_colors 生成了色彩图:

```
>>> palette = sns.color_palette('viridis', n_colors=6)
```

重新绘制的结果如图 6.40 所示,使单独着色的线条更加明显,特别是线宽稍粗的线条。

```
>>> sns.relplot(x='time', y='firing_rate',palette=palette,
...             hue='coherence', style='choice',
...             kind='line', data=dots, linewidth=2)
<seaborn.axisgrid.FacetGrid object at 0x7f937085a130>
```

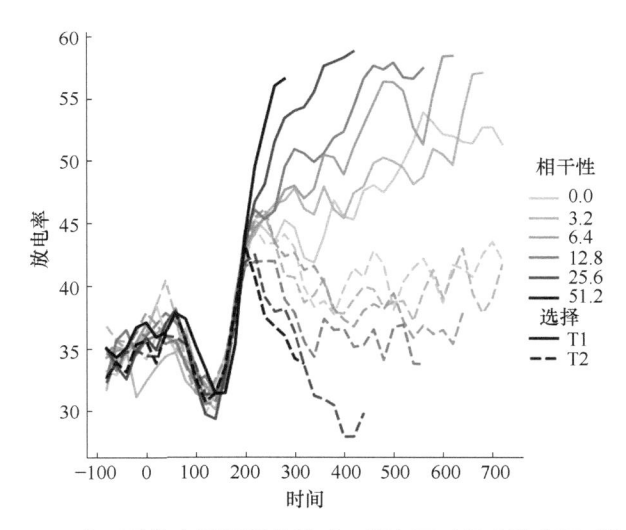

图 6.39 当列是数字而不是分类时,颜色可以沿连续范围列缩放

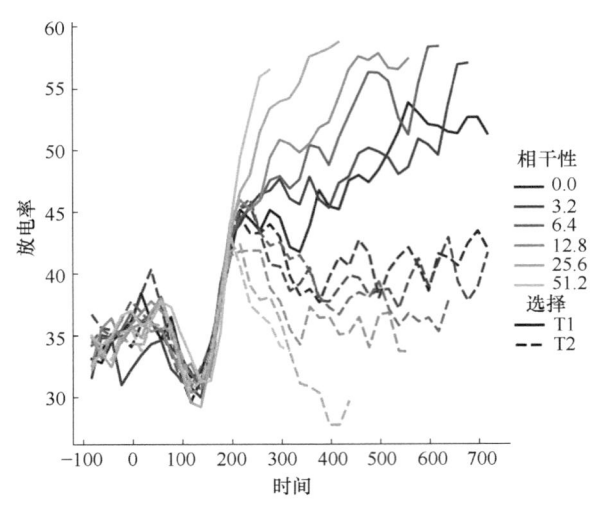

图 6.40　色彩图

6.2.2　多个绘图

由于 relplot 返回 FacetGrid 对象，因此通过指定 row 和 col 关键字参数来创建多个子图是很简单的。下面将两个子图并排堆叠在两列中，因为数据框中的 time 列有两个唯一值，如图 6.41 所示。

```
>>> tips.time.unique()
['Dinner', 'Lunch']
Categories (2, object): ['Dinner', 'Lunch']
>>> sns.relplot(x='total_bill', y='tip', hue='smoker',
...             col='time',  # 时间列
...             data=tips)
<seaborn.axisgrid.FacetGrid object at 0x7f9370704ee0>
```

可以沿着多行多列生成多个网格图。col_wrap 关键字参数可以防止图 6.42 填充过宽，而是将每个子图放入单独的行中。

```
>>> sns.relplot(x='timepoint', y='signal', hue='event',
...             style='event',
...             col='subject', col_wrap=5,
...             height=3, aspect=.75, linewidth=2.5,
...             kind='line',
...             data=fmri.query('region == "frontal"'))
<seaborn.axisgrid.FacetGrid object at 0x7f9370895be0>
```

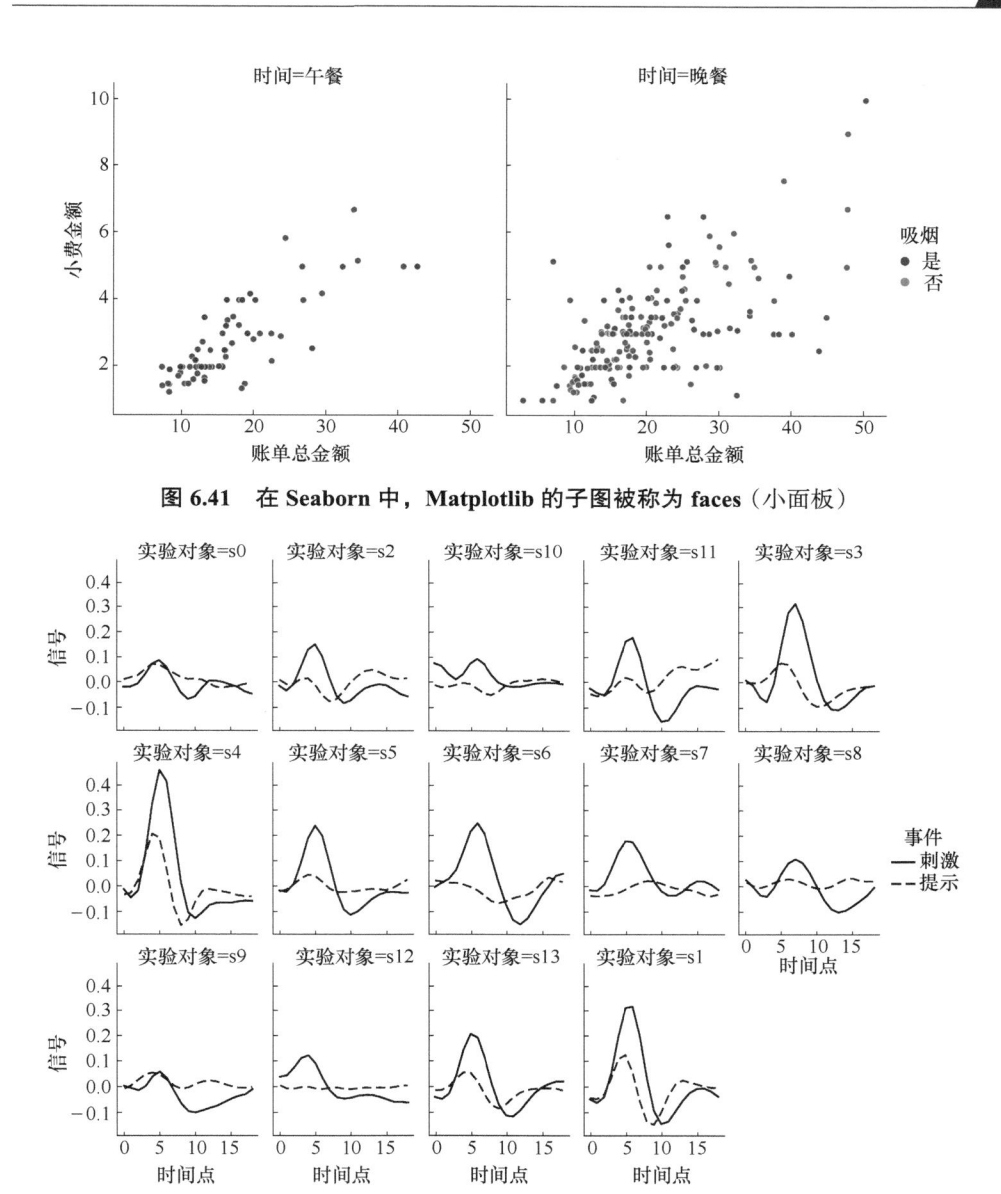

图 6.41 在 Seaborn 中，Matplotlib 的子图被称为 faces（小面板）

图 6.42 Faces 支持复杂的图形布局

6.2.3 分布图

可视化数据分布对于任何统计分析以及诊断机器学习模型问题都是至关重要的。Seaborn 提供的可视化数据分布功能非常强大。最简单的数据分布可视化方法是直方图。

```
>>> penguins = sns.load_dataset("penguins")
>>> penguins.head()
  species     island  bill_length_mm  bill_depth_mm  flipper_length_mm  body_mass_g     sex
0  Adelie  Torgersen           39.10          18.70             181.00      3750.00    Male
1  Adelie  Torgersen           39.50          17.40             186.00      3800.00  Female
2  Adelie  Torgersen           40.30          18.00             195.00      3250.00  Female
3  Adelie  Torgersen             nan            nan                nan          nan     NaN
4  Adelie  Torgersen           36.70          19.30             193.00      3450.00  Female
```

Seaborn 的 displot 可以快速绘制直方图，如图 6.43 所示。

```
>>> sns.displot(penguins, x="flipper_length_mm")
<seaborn.axisgrid.FacetGrid object at 0x7f9370951b20>
```

直方图中的 bins 的宽度可以使用 binwidth 关键字参数进行选择，bins 的数量可以使用 bins 关键字参数进行选择。对于具有少量不同值的分类数据，bins 可以选择为一系列不同值，例如 bins=[1, 3, 5, 8]，或者使用 discrete=True 关键字参数自动处理这个问题。

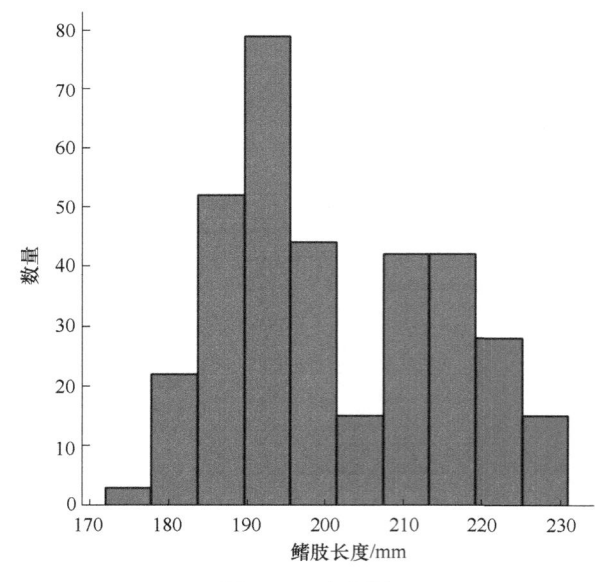

图 6.43 直方图

还可以创建重叠的半透明直方图，如图 6.44 所示。

```
>>> sns.displot(penguins, x="flipper_length_mm", hue="species")
<seaborn.axisgrid.FacetGrid object at 0x7f9363c04f70>
```

Seaborn 智能地为每个直方图选择要匹配的 bins，而 Matplotlib 的默认 hist（）函数很难做到这一点。

有时，直方图的垂直线可能会分散注意力，可以使用 element='step' 关键字参数删除它，如图 6.45 所示。

```
>>> sns.displot(penguins, x='flipper_length_mm',
...                       hue='species',
...                       element='step')
<seaborn.axisgrid.FacetGrid object at 0x7f9363b01f70>
```

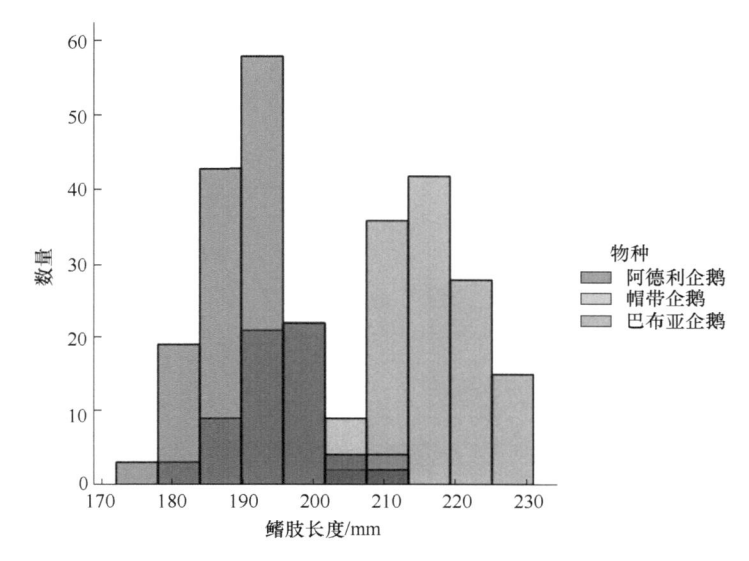

图 6.44 可以使用颜色和透明度叠加多个直方图

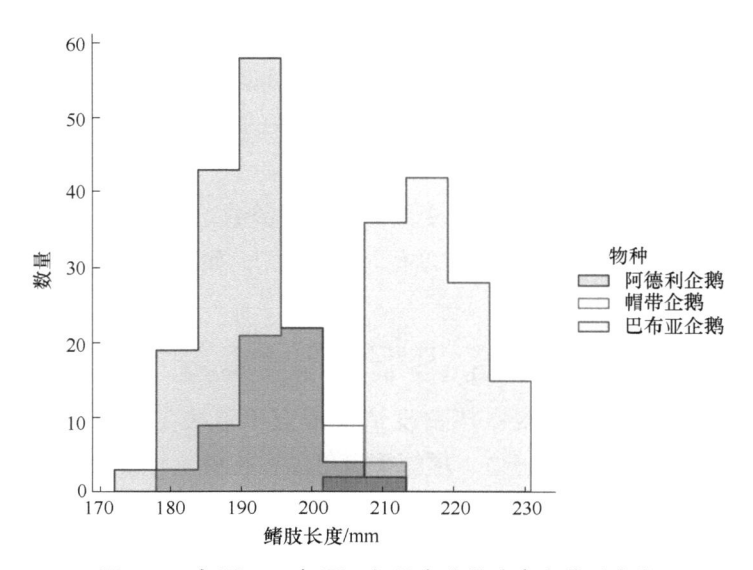

图 6.45 与图 6.44 相同，但没有分散注意力的垂直线

除了使用透明度进行层叠，直方图可以相互堆叠，如图 6.46 所示。但如果堆叠的数量过多，则一些主要的直方图可能会掩盖其他直方图。

```
>>> sns.displot(penguins, x="flipper_length_mm",
...                       hue="species",
...                       multiple="stack")
<seaborn.axisgrid.FacetGrid object at 0x7f9372f6fbb0>
```

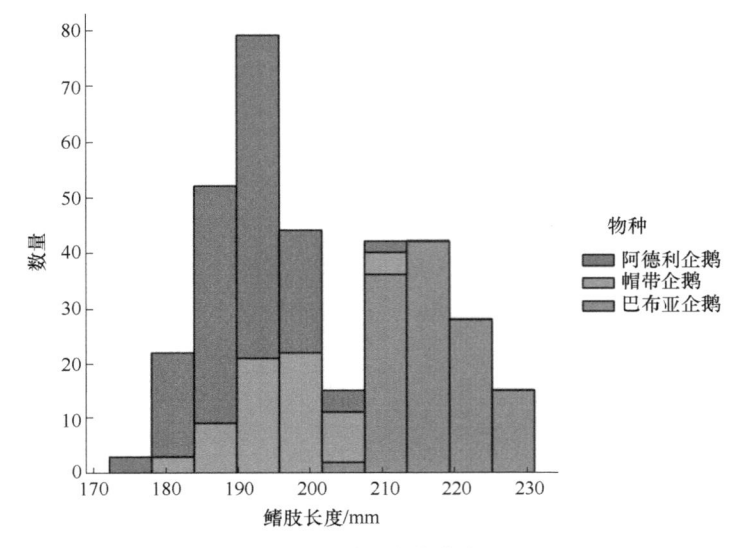

图 6.46　叠加多个直方图

如果有多个直方图，可以使用 multiple='dodge' 关键字参数将柱状变窄以适应并排显示。如果要将直方图用作概率密度函数的估计，则必须适当进行缩放，这可以通过 stat='probability' 关键字参数来实现。

直方图使用矩形函数来近似单变量概率密度函数，但是可以使用高斯函数通过 kind 关键字参数创建平滑的核密度估计（KDE），如图 6.47 中所示的相同概率密度函数。

```
>>> sns.displot(penguins, x="flipper_length_mm", kind="kde")
<seaborn.axisgrid.FacetGrid object at 0x7f936398f3d0>
```

生成的图表的平滑程度由核密度估计带宽参数决定，该参数可作为 bw_adjust 关键字参数提供。请记住，尽管平滑可能在视觉上更美观，但可能会误传不连续性或抑制数据中可能重要的特征。

除了单变量概率密度函数之外，Seaborn 还为双变量概率密度函数提供了强大的可视化工具（见图 6.48）。

```
>>> sns.displot(penguins, x="bill_length_mm", y="bill_depth_mm")
<seaborn.axisgrid.FacetGrid object at 0x7f9363635730>
```

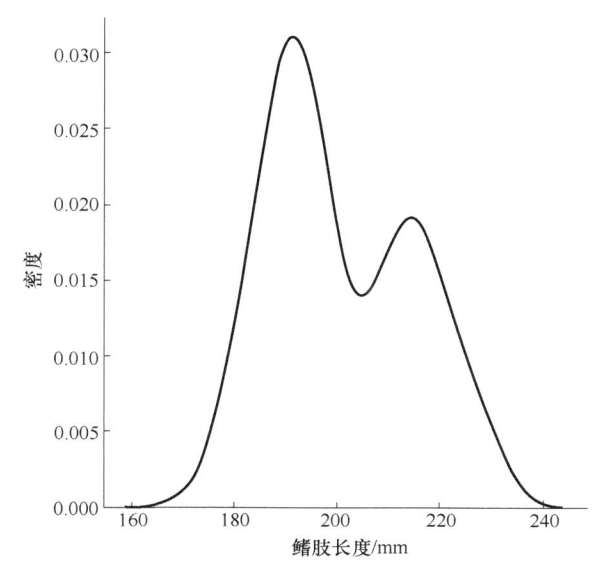

图 6.47　Seaborn 支持核密度估计

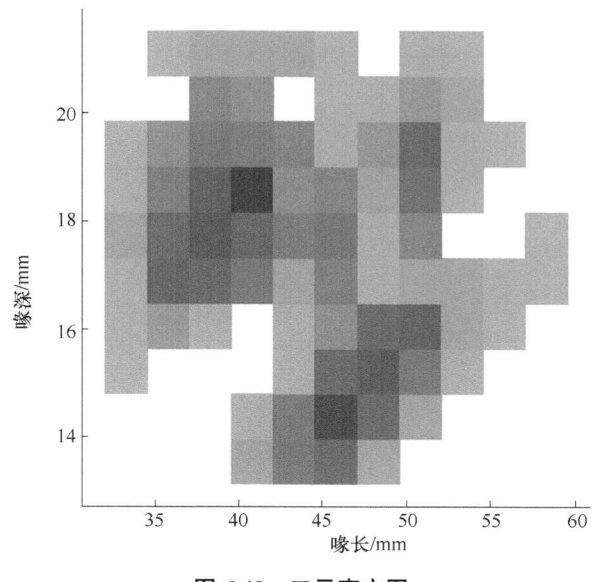

图 6.48　二元直方图

图 6.48 显示了二维网格，该网格将每个数据点的数量与这些计数的相应色阶相匹配。使用 kind='kde' 关键字参数也适用于双变量分布。如果双变量分

布之间没有太多重叠，则可以使用 hue 关键字参数以不同的颜色绘制它们（见图 6.49）。可以使用 binwidth 关键字参数为每个坐标维度选择不同的 bin 宽度，但需要一个元组来定义每个坐标维度的 bin 宽度。cbar 关键字参数绘制相应的色阶，但当与 hue 关键字参数组合时，它将绘制多个色阶（每个色阶对应一个用 hue 编码的类别），这使得生成的图 6.49 非常拥挤。

```
>>> sns.displot(penguins, x='bill_length_mm',
...             y='bill_depth_mm', hue='species')
<seaborn.axisgrid.FacetGrid object at 0x7f93632f2040>
```

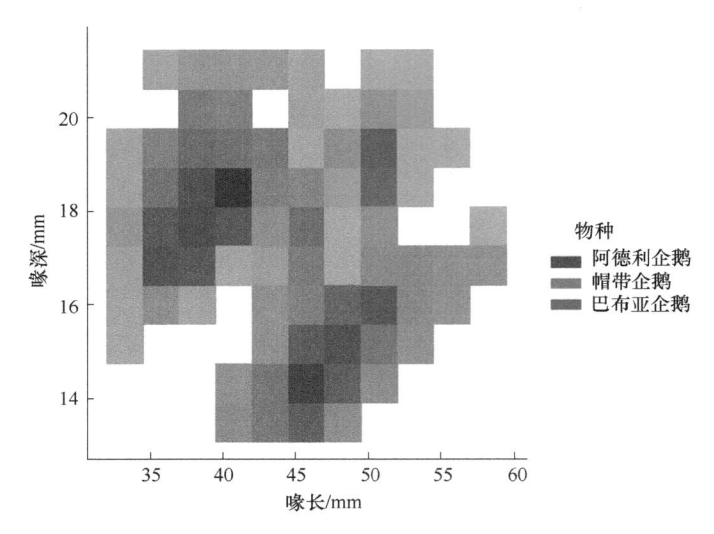

图 6.49　多个二元直方图可以用不同的颜色绘制在一起

对绘制双变量分布的边缘分布非常有帮助，Seaborn 使用 jointplot（见图 6.50）可以轻松地实现这一点。

```
>>> sns.jointplot(data=penguins,
...               x="bill_length_mm",
...               y="bill_depth_mm")
<seaborn.axisgrid.JointGrid object at 0x7f9362f25670>
```

jointplot 返回一个 JointGrid 对象，该对象允许使用 plot_marginals（）方法单独绘制边距，如图 6.51 所示。

```
>>> g = sns.JointGrid(data=penguins,
...                   x="bill_length_mm",
...                   y="bill_depth_mm")
>>> g.plot_joint(sns.histplot)
<seaborn.axisgrid.JointGrid object at 0x7f93632e8f10>
>>> g.plot_marginals(sns.kdeplot,fill=True)
<seaborn.axisgrid.JointGrid object at 0x7f93632e8f10>
```

注意，边缘分布现在是核密度估计图而不是 bin 线图。plot_marginals（）中未使用的关键字参数将传递给函数参数（在这种情况下为 fill=True，用于 sns.kdeplot）。除了 jointplot，Seaborn 还提供了 pairplot，该函数会为输入数据框中所有列的每一对数据生成一个 jointplot。

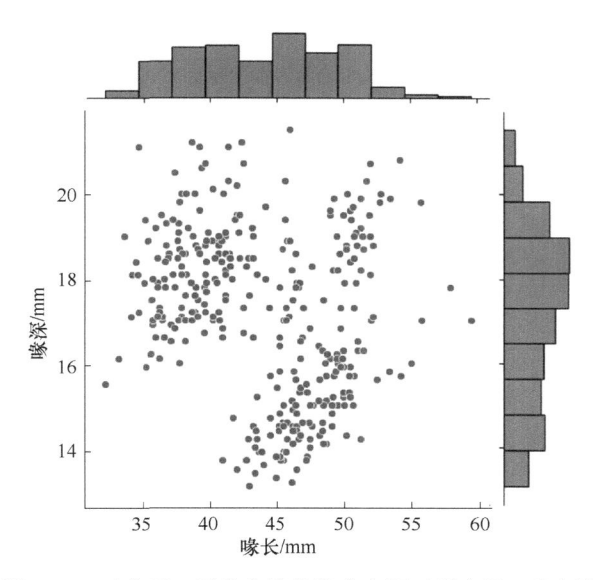

图 6.50　对应于二元分布的边缘分布可以沿水平 / 垂直轴

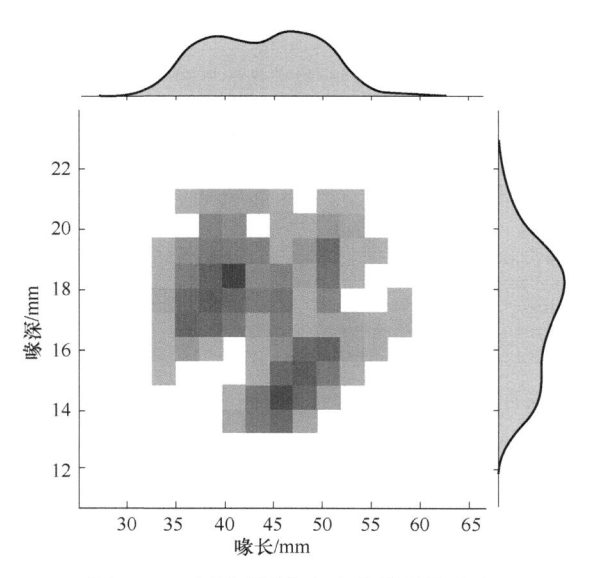

图 6.51　也可以画出自定义的边缘分布

Seaborn 提供了许多选项用于可视化分类数据。虽然我们主要使用类别来提供不同的颜色或叠加图形，但 Seaborn 提供了专门针对分类数据的可视化方法。例如，图 6.52 中显示了一周中每天的 total_bill 的散点图。

```
>>> sns.catplot(x="day", y="total_bill", data=tips)
<seaborn.axisgrid.FacetGrid object at 0x7f93623c2610>
```

重要的是，个别点沿水平方向随机分散，以避免叠加和遮挡点。可以使用 jitter=False 关键字参数关闭此功能。

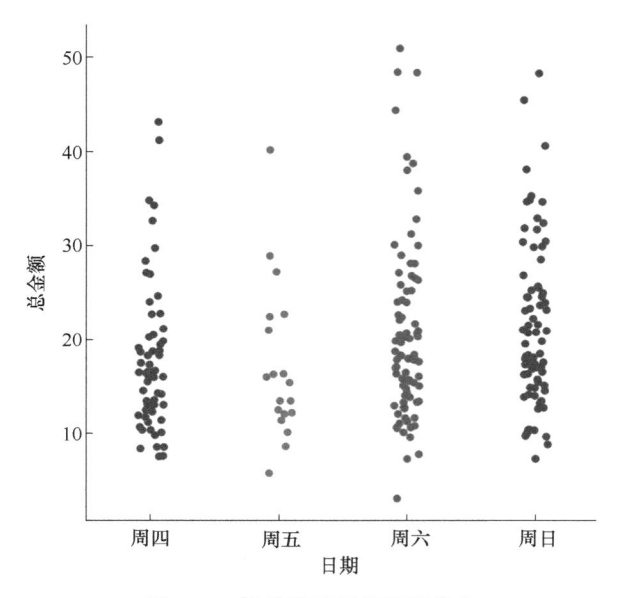

图 6.52 沿分类变量的离散分布

另一种有趣的方法是通过使用 kind='swarm' 关键字参数来使用 swarmplot，而不是使用抖动来避免模糊数据，如图 6.53 所示。

```
>>> sns.catplot(data=tips,
...             x="day",
...             y="total_bill",
...             hue="sex",
...             kind="swarm")
<seaborn.axisgrid.FacetGrid object at 0x7f93623cb220>
```

在图 6.53 中，采用类似扩展的圣诞树树枝的范围取代了由于抖动而产生的随机定位。这种图的一个迷人之处是它产生的聚类效应，它会自动将你的注意力吸引到某些事先很难确定的特征上。例如，图 6.53 似乎显示周六男性的总账单高于女性，特别是在 total_bill = 20 左右。使用 Seaborn 的 displot，我们可以快速验证，下面是图 6.54 的代码：

```
>>> sns.displot(tips.query('day=="Sat" and sex=="Male"'),
...             x='total_bill')
<seaborn.axisgrid.FacetGrid object at 0x7f9362021a90>
```

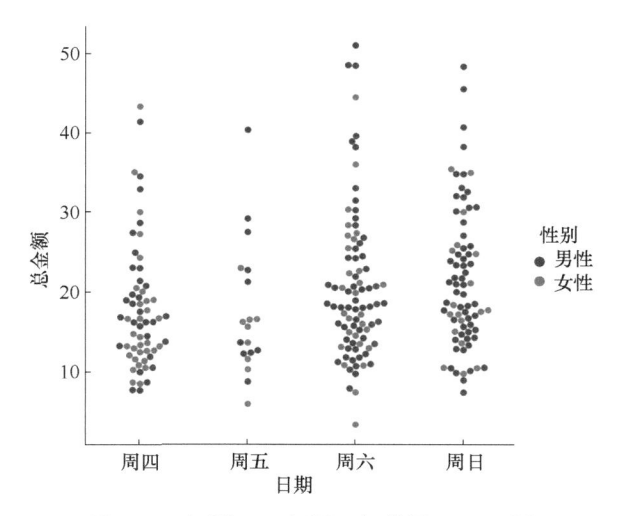

图 6.53　与图 6.52 相同，但使用 swarm 图

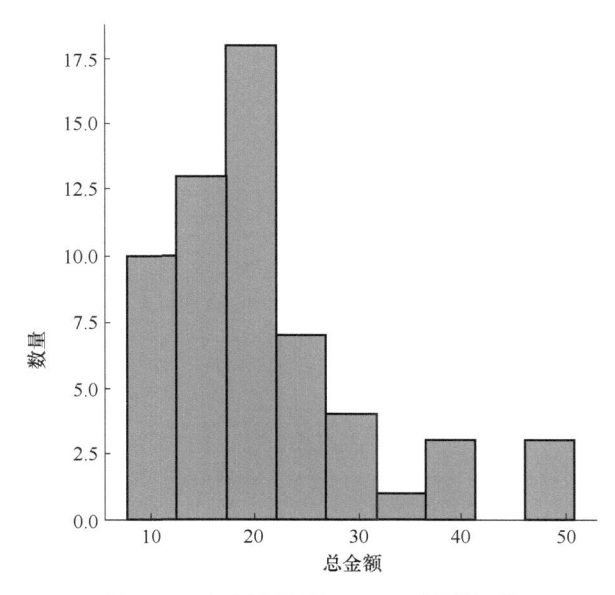

图 6.54　直方图显示了 swarm 图的细节

Seaborn 图形的颜色、线条和整体呈现方式可以通过 sns.set_theme（ ）进行全局控制，或者通过使用上下文管理器（例如 with sns.axes_style("white")）在特定图形级别进行控制。通过 set（ ）函数可以更精细地控制字体和其他细节。

Seaborn 支持离散数据的颜色序列或连续数据的明显区分色，正如我们之前在 color_palette 中看到的那样。人类的感知系统对颜色的反应非常敏感，Seaborn 在管理数据颜色时已经解决了许多常见的视觉陷阱。

6.3 Bokeh

基于 Web 的可视化技术和 Python 的结合为利用现代 JavaScript 框架进行基于网络的交互式可视化开辟了广泛的可能性。这使得部署 Python 数据可视化成为可能，用户可以通过 Web 浏览器如 Google Chrome 或 Mozilla Firefox（而不是 Internet Explorer）进行交互。Bokeh 是一个开源的 Python 模块，它使开发和部署这种基于 Web 的交互式可视化变得更加容易，因为它提供一些 JavaScript 的基本组件。

6.3.1 使用 Bokeh 基元

下面的代码生成了图 6.55：

```
from bokeh.plotting import figure, output_file, show
# 数据准备
x = range(10)
y = [i**2 for i in x]
# 创建一个带有标题和轴标签的新图形
p = figure(title="plot the square", # 图题
           x_axis_label='x',
           y_axis_label='y',
           width= 400, # 图宽
           height = 300) # 图高
# 添加带有图例和线条粗细的线条渲染器
p.line(x, y,
       legend=r"x^2", # 图例中显示的文本
       line_width=2) # 线条宽度

# 显示结果
show(p)
```

Bokeh 与 Matplotlib 遵循相同的图层分解理念。由于 Bokeh 可以渲染到多个终端（包括 Jupyter Notebook），因此我们使用 output_file 函数将输出设置为 HTML 文件。Bokeh 绘图的主要步骤是使用 figure 函数创建图形画布，并为图形提供参数，例如 x_axis_label 和 width。然后使用 line 方法在 figure 对象上添加线条，并包括线宽等线条参数。最后，show 函数实际上将 HTML 写入 output_file

中指定的文件。因此，Bokeh 的关键优势在于它允许在 Python 中进行交互式可视化的分阶段，并将必要的 JavaScript 渲染到 HTML 输出文件中。

在图形的右边空白处，我们有图形的默认工具。顶部的 pan 工具可以在图形框架中拖动线条，紧接其下方是 zoom 工具。这些工具以 JavaScript 函数的形式嵌入 HTML 中，并由浏览器运行。这意味着 Bokeh 同时创建了静态 HTML 和嵌入式的 JavaScript。同样，绘图的数据也以类似的方式嵌入生成的 HTML 文档中。与 Matplotlib 一样，Bokeh 可以对画布上的每个元素进行低级别的控制。同时，这意味着你可以使用工具包中的基本组件来创建功能强大、引人入胜的图形，但代价是需要对图形中的每个元素进行编程。像 Matplotlib 一样，Bokeh 还提供一个详细的示例图库[一]，你可以将其作为自己可视化的起点。

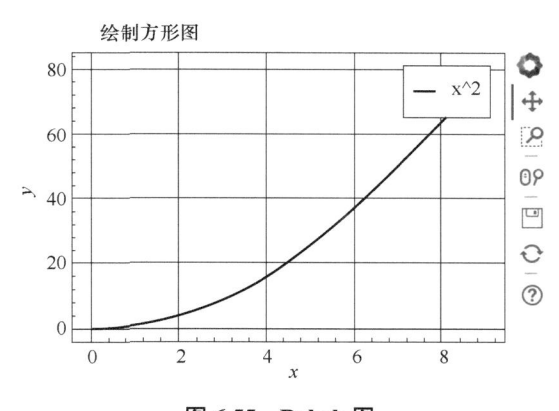

图 6.55　Bokeh 图

当将图形元素（即 glyph）添加到画布时，可以指定其各个属性。下一行代码添加了指定半径的红色圆圈，以标记数据点中的每个点（见图 6.56）。

```
p.circle(x,y,radius=0.2,fill_color='red')
```

或者对单个对象添加属性：

```
c = p.circle(x,y)
c.glyph.radius= 0.2
c.glyph.fill_color = 'red'
```

要向图形添加多个图形元素，可以使用图形对象的方法（例如，p.circle（），p.line（），p.square（））。参考文献［1］中列出了可用基本图形元素的详尽清单，只需学习可用图形元素的术语并分配其属性即可。

[一] https://docs.bokeh.org/en/latest/docs/gallery.html。

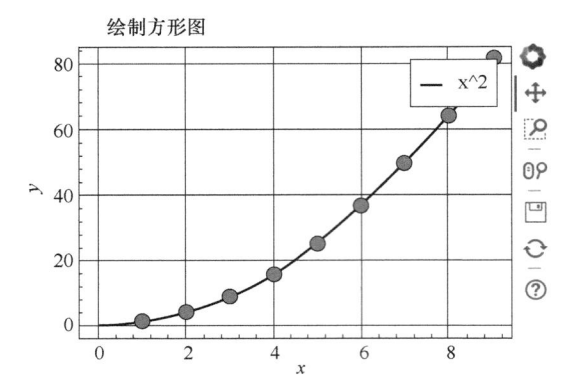

图 6.56　绘制自定义标记数据点

6.3.2　Bokeh 布局

类似于 Matplotlib 中的 subplot 函数，Bokeh 具有 row 和 column 函数，可将单独的图形排列在同一画布上。下面的代码使用 row 函数将两个图形并排放置（见图 6.57）：

```
from bokeh.plotting import figure, show
from bokeh.layouts import row, column
from bokeh.io import output_file
output_file('bokeh_row_plot.html')
x = range(10)
y = [i**2 for i in x]
f1 = figure(width=200,height=200)
f1.line(x,y,line_width=3)
f2 = figure(width=200,height=200)
f2.line(x,x,line_width=4,line_color='red')
show(row(f1,f2))
```

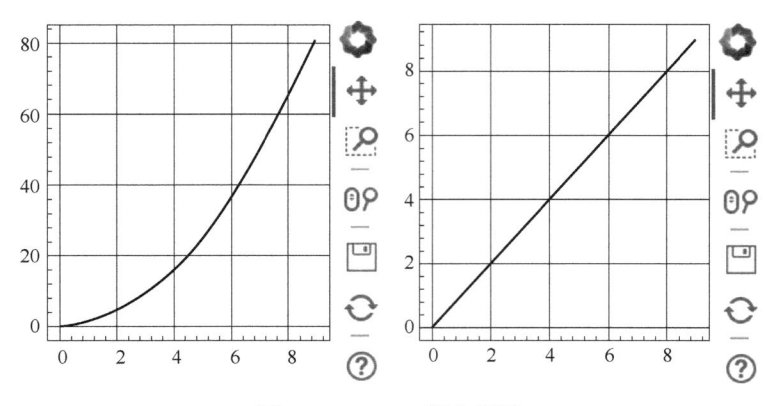

图 6.57　Bokeh 行布局图

我们创建了各个图形对象，然后使用 row（ ）函数的 show（ ）方法将两个图形 f1 和 f2 并排显示。我们还可以将其与 column 函数结合使用，以创建网格布局（见图 6.58）。

```python
from bokeh.plotting import figure, show
from bokeh.layouts import row, column
from bokeh.io import output_file

output_file('bokeh_row_column_plot.html')

x = range(10)
y = [i**2 for i in x]
f1 = figure(width=200,height=200)
f1.line(x,y,line_width=3)
f2 = figure(width=200,height=200)
f2.line(x,x,line_width=4,line_color='red')
f3 = figure(width=200,height=200)
f3.line(x,x,line_width=4,line_color='black')
f4 = figure(width=200,height=200)
f4.line(x,x,line_width=4,line_color='green')
show(column(row(f1,f2),row(f3,f4)))
```

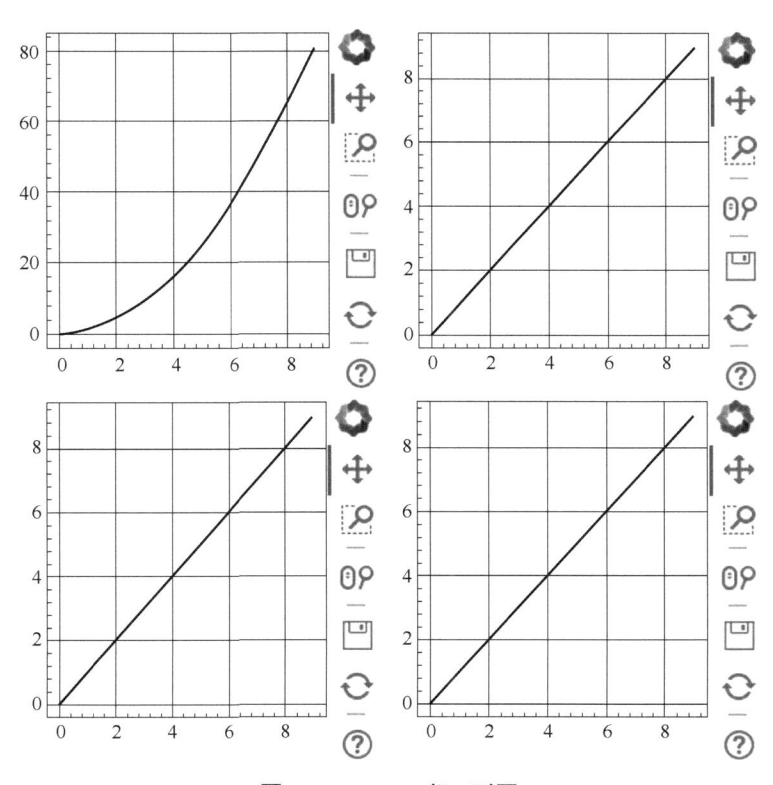

图 6.58　Bokeh 行 - 列图

注意，由于各个图形是分开处理的，因此它们有自己的工具组。这可以通过使用 bokeh.layouts.gridplot 或更通用的 bokeh.layouts.layout 函数来解决。

6.3.3　Bokeh 组件

Bokeh 交互式组件允许用户进行可视化交互和探索，这些小组件主要基于 Bootstrap JavaScript 库。所有小组件都遵循相同的两步实现模式。第一步是创建小组件，将其布局在画布上。第二步是创建回调函数，当用户与小组件进行交互时将触发该回调函数。

假设回调指定了基于小组件的某种操作，那么该操作将在哪里发生呢？如果输出是由浏览器渲染的 HTML 文件，那么该操作必须通过 JavaScript 处理并在浏览器中运行。当交互由 Python 工作空间中的对象驱动或依赖于这些对象时，问题就会变得复杂起来。请记住，在创建 HTML 文件之后，Python 就不再起作用了。如果你希望利用 Python 进程的回调函数，则必须使用 bokeh serve 来托管该应用程序（见图 6.59）。

让我们从由 JavaScript 处理的回调函数开始，这些回调函数在静态 HTML 输出中进行处理，就像下面的示例中一样：

```
from bokeh.io import output_file, show
from bokeh import events
from bokeh.models.widgets import Button
from bokeh.models.callbacks import CustomJS
output_file("button.html")
cb = CustomJS(args=dict(),code='''
    alert("ouch!");
    ''')
button = Button(label='Hit me!') # 创建 button 对象
button.js_on_event(events.ButtonClick, cb)
show(button)
```

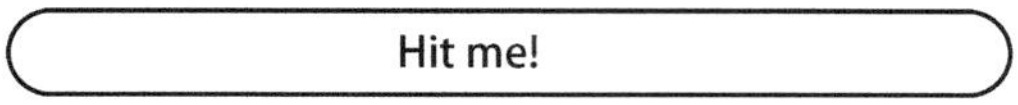

图 6.59　按钮小组件的 JavaScript 回调

这里的关键步骤在于 CustomJS 函数，该函数获取有效 JavaScript 的嵌入字符串并将其打包为静态 HTML 文件。按钮小组件没有任何参数，因此 args 变量只是一个空字典。下一个重要部分是 js_on_event 函数，它指定事件（即 ButtonClick）和分配给处理该事件的回调函数。现在，当创建输出 HTML 文件

第 6 章　可视化数据

并在浏览器中渲染页面，然后单击名为"Hit me！"的按钮时，你将在浏览器中收到一个弹出窗口，其中包含文本"ouch"。

根据这个结构，我们可以尝试一些更复杂的东西，如下所示（见图 6.60）：

```
from bokeh.io import output_file, show
from bokeh.models.callbacks import  CustomJS
from bokeh.layouts import widgetbox
from bokeh.models.widgets import Dropdown
output_file("bokeh_dropdown.html")
cb = CustomJS(args=dict(),code='alert(cb_obj.value)')
menu = [("Banana", "item_1"),
        ("Apple", "item_2"),
        ("Mango", "item_3")]
dropdown = Dropdown(label="Dropdown button", menu=menu,callback=cb)
show(widgetbox(dropdown))
```

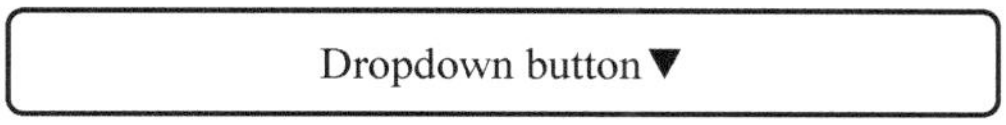

Dropdown button ▼

图 6.60　下拉小组件的 JavaScript 回调

在这个例子中，我们使用了 DropDown 小组件，并用菜单中的项目填充它。元组的第一个元素是下拉菜单中显示的字符串，第二个元素是与之关联的 JavaScript 变量。回调函数与之前相同，只是现在加入了 cb_obj 变量。这是在激活时自动传递到 JavaScript 函数中的回调对象。在这种情况下，我们在回调关键字参数中指定了回调，而不是像以前那样依赖于事件类型。现在，页面在浏览器中渲染后，当你在菜单中的某个元素上单击并下拉时，应该能看到弹出窗口。

下面的示例结合了我们之前创建的下拉菜单和折线图。这里唯一的新元素是通过 args 关键字参数将线对象传递到 CustomJS 函数中。一旦进入 JavaScript 代码，我们就可以根据下拉菜单中选择的值通过 cb_obj.value 变量来更改线条图元的嵌入线条颜色。这使我们能够通过小组件控制线条的颜色属性。使用相同的方法，只要通过 args 关键字参数将它们传递给 CustomJS 函数，我们就可以改变其他对象的其他属性（见图 6.61）。

```
from bokeh.plotting import figure, output_file, show
from bokeh.layouts import column
from bokeh.models.widgets import Dropdown
from bokeh.models.callbacks import  CustomJS
output_file("bokeh_line_dropdown.html")
# 创建数据
x = range(10)
y = [i**2 for i in x]
# 创建 figure
p = figure(width = 300, height=200, # 长宽设定
           tools="save,pan", # 工具组件
```

195

```python
                x_axis_label='x',    # 坐标轴
                y_axis_label='y')
# 添加行
line = p.line(x,y,                   # x,y 数据
              line_color='red',      # 红色
              line_width=3)          # 线宽
# 下拉组件元素
menu = [("Red", "red"),("Green", "green"),("Blue", "blue")]
# 传递 line 对象并根据下拉选择更改 line_color
cb = CustomJS(args=dict(line=line),
              code='''
                   var line = line.glyph;
                   var f = cb_obj.value;
                   line.line_color = f;
                   ''')
# 将回调分配给下拉组件
dropdown = Dropdown(label="Select Line Color",
↪    menu=menu,callback=cb)
# Dropdown 置于图片之上
show(column(dropdown,p))
```

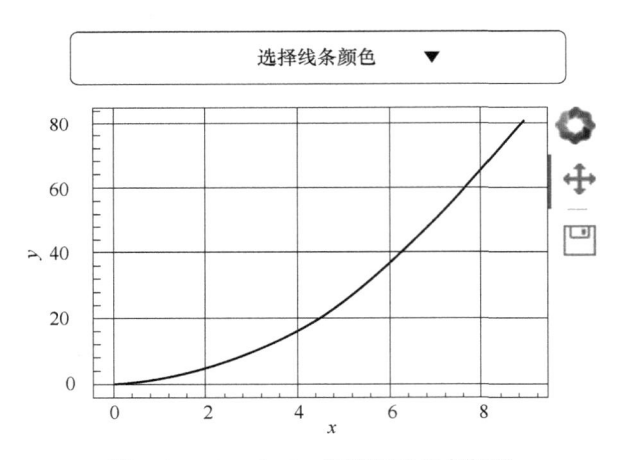

图 6.61　JavaScript 回调更改线条颜色

最后一个示例展示了如何使用浏览器执行的回调函数来操作线条属性。Bokeh 可以通过更改数据本身并让嵌入的图形对这些更改做出反应。下面的示例与上一个示例非常相似，只是这里我们使用 ColumnDataSource 对象将数据从 Python 传输到浏览器。然后，在嵌入的 JavaScript 代码中，我们解包数据，再根据下拉菜单中的操作更改数据数组。关键部分是触发数据更新并绘制，使用 source.change.emit（）函数（见图 6.62）。

```python
from bokeh.plotting import figure, output_file, show
from bokeh.layouts import row, column
from bokeh.models import ColumnDataSource
from bokeh.models.widgets import Dropdown
from bokeh.models.callbacks import import  CustomJS
```

```python
import numpy as np
output_file("bokeh_ColumnDataSource.html")
# 创建数据
t = np.linspace(0,1,150)
y = np.cos(2*np.pi*t)
# 创建 ColumnDataSource 并打包数据
source = ColumnDataSource(data=dict(t=t, y=y))
# 创建 figure
p = figure(width = 300, height=200, tools="save,pan",
           x_axis_label='time (s)',y_axis_label='Amplitude')
# 增加 line,但现在使用 ColumnDataSource
line = p.line('t','y',source=source,line_color='red')
menu = [("1 Hz", "1"), ("5 Hz", "5"), ("10 Hz", "10")]
cb = CustomJS(args=dict(source=source),
              code='''
                    var data = source.data;
                    var f = cb_obj.value;
                    var pi = Math.PI;
                    t = data['t'];
                    y = data['y'];
                    for (i = 0; i < t.length; i++) {
                        y[i] = Math.cos(2*pi*t[i]*f)
                    }
                    source.change.emit();
                    ''')
dropdown = Dropdown(label="Select Wave Frequency",
↪    menu=menu,callback=cb)
show(column(dropdown,p))
```

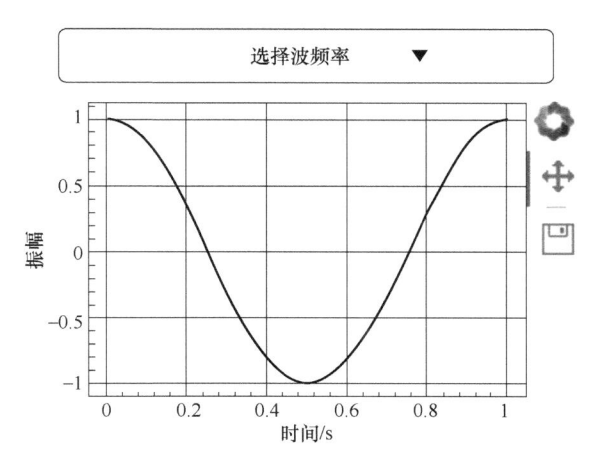

图 6.62 JavaScript 回调更新图形正弦波频率

这种方法的优点是，一方面它创建了一个自包含的 HTML 文件，不再需要 Python。另一方面，回调可能需要主动的 Python 计算，而这对于浏览器中的 JavaScript 来说是不可能的。在这种情况下，我们可以使用 bokeh server，如下所示：

```
Terminal> bokeh serve bokeh_ColumnDataSource_server.py
```

它将运行一个小型服务器并打开一个指向本地页面的网页。该网页基本上包含与上一个示例相同的内容,尽管你可能会注意到该页面的响应性较上一个示例要差。这是因为交互必须往返于服务器进程和浏览器之间,而不是在浏览器本身中更新。bokeh_ColumnDataSource_server 的内容如下:

```python
from bokeh.plotting import figure, show, curdoc
from bokeh.layouts import column
from bokeh.models import ColumnDataSource, Select
from bokeh.models.widgets import Dropdown
import numpy as np
# 创建数据
t = np.linspace(0,1,150)
y = np.cos(2*np.pi*t)
# 创建 ColumnDataSource 并打包数据
source = ColumnDataSource(data=dict(t=t,y=y))

# 创建 figure
p = figure(width = 300, height=200, tools="save,pan",
           x_axis_label='time (s)',y_axis_label='Amplitude')
# 增加 line,但现在使用 ColumnDataSource
line = p.line('t','y',source=source,line_color='red')
menu = [("1 Hz", "1"),
        ("5 Hz", "5"),
        ("10 Hz", "10")]
def cb(attr,old,new):
    f=float(freq_select.value)
    d = dict(x=source.data['t'],y=np.cos(2*np.pi*t*f))
    source.data.update(d)

freq_select = Select(value='1 Hz', title='Frequency (Hz)',
options=['1','5','10'])
freq_select.on_change('value',cb)
curdoc().add_root(column(freq_select,p))
```

注意,上面的代码中 Select 取代了 Dropdown。freq_select.on_change('value', cb)是下拉选择与 Python 服务器进程通信的方式。重要的是,options 关键字参数接受字符串序列['1','5','10'],即使我们必须在回调中将字符串转换回浮点数。

```python
freq_select = Select(value='1 Hz',
                     title='Frequency (Hz)',
                     options=['1','5','10'])
```

代码中的最后一个关键步骤是 curdoc( ).add_root(column(freq_select, p)),它创建了一个与文档关联并响应回调的定向列,并将其附加到文档中,这样它就能响应回调。注意,部署 Bokeh 服务器有许多配置选项,包括使用 Tornado 作为后端并运行多个线程(有关详细信息,请参见主文档)。Bokeh 还与 Jupyter Notebook 有许多集成选项。

Bokeh 处于积极开发中,其整体设计和架构都经过了充分的考虑。通过 Web

浏览器提供的 JavaScript 和 Python 可视化的结合，是开源社区中一个快速发展的领域。Bokeh 只是这一联合策略的众多实现之一（并且可以说是最优秀的实现之一）。请关注 Bokeh 团队的进一步开发和新特性，但记住，与 Matplotlib 相比，Bokeh 的成熟度较低。

6.4 Altair

在基于 Web 的科学可视化领域的另一个项目是 Altair，它是一个声明式可视化模块。与 Matplotlib 需要指定构建的所有细节（即，命令式可视化）不同，Altair 实现了用于可视化的 Vega-Lite JSON 图形语法。这意味着底层呈现的图形实际上是通过 Vega 在 JavaScript 中实现的。与 Altair 一起工作的最佳方式是通过 Jupyter Notebook，在使用 Jupyter Notebook 时可能需要启用 Altair 渲染器。

```
import altair as alt
```

Altair 的主要对象是 Chart 对象。

```
>>> import altair as alt
>>> from altair import Chart
>>> import vega_datasets
>>> cars = vega_datasets.data('cars')
>>> chart = Chart(cars)
```

重要的是，输入 Chart 的 data 应该是一个 Pandas Dataframe、altair.Data 对象，或者是引用 csv 或 JSON 文件（采用所谓的整洁模式）的 URL。整洁格式基本上意味着 dataframe 的列是变量，行是这些变量的观测值。要创建 Altair 可视化，必须决定数据值的标记（即字形）。执行 chart.mark_point() 只会在 Jupyter Notebook 中绘制一个圆圈。要绘制图，你必须为 Dataframe 的其他元素指定通道（见图 6.63）。

```
>>> chart.mark_point().encode(x='Displacement')
alt.Chart(...)
```

它使用输入数据框的 Displacement 列创建了一个一维散点图。在概念上，encode 意味着在数据和其视觉表示

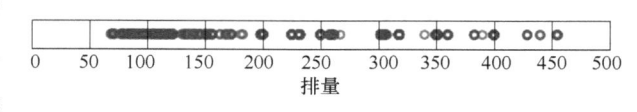

图 6.63　一维 Altair 图

之间创建映射。使用这种模式，要创建一个二维 X-Y 图，我们需要为 y 数据指定一个额外的通道，如下所示（见图 6.64）：

```
>>> chart.mark_point().encode(x = 'Displacement',
...                           y = 'Acceleration')
alt.Chart(...)
```

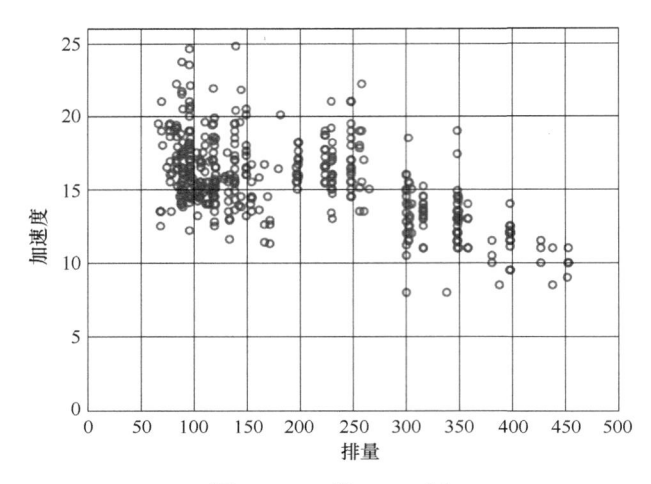

图 6.64　二维 Altair 图

这将创建一个加速度与位移的二维图。注意，图表对象可以访问输入数据框中的列名称，因此我们可以通过它们的字符串名称作为关键字参数传递给 encode 方法来访问它们。由于每行都有一个对应的分类名称，我们可以将其用作在所创建的 X-Y 图中的颜色维度，使用 color 关键字参数如下（见图 6.65）：

```
>>> chart.mark_point().encode(x='Displacement',
...                           y='Acceleration',
...                           color='Origin')
alt.Chart(...)
```

使用 Altair 的关键是要认识到它是一个利用 Vega-Lite JavaScript 可视化库的轻量级层。因此，你可以在 Altair 中构建任何类型的 Vega-Lite 可视化，并将其连接到 Pandas 数据框或其他 Python 构造，而无需自己编写 JavaScript。如果你想要更改图形的个别颜色或线条宽度，那么需要使用 Vega 编辑器，然后将 Vega-Lite 规范重新引入 Altair，这并不困难，但涉及的步骤较多。关键是，这种定制化并不是 Altair 的强项。作为一个声明式的可视化模块，其目的是将这些细节留给 Altair 处理，而将注意力集中在将数据映射到可视元素上。

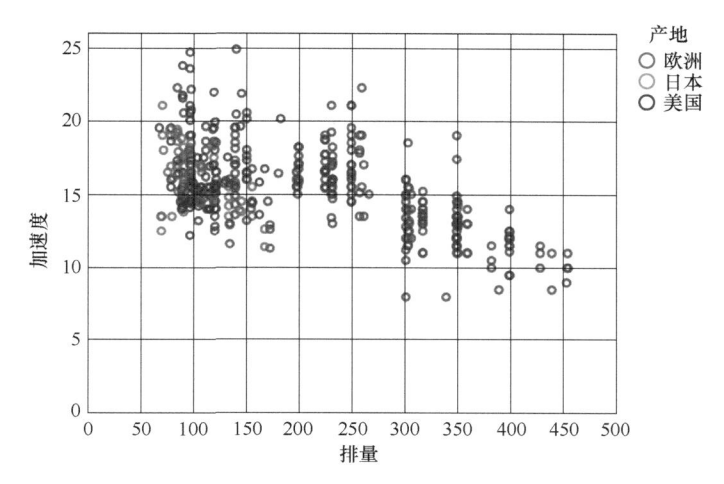

图 6.65 每个标记的颜色由输入 Dataframe 中的 Origin（产地）列确定

6.4.1 Altair 细节化

在 Altair 中控制视觉呈现的各个元素可以通过使用一系列 configure_ 顶层函数来实现，如图 6.66 所示。

```
>>> (chart.configure_axis(titleFontSize=20,
...                       titleFont='Consolas')
...       .mark_point()
...       .encode(x='Displacement',
...               y='Acceleration',
...               color='Origin')
... )
alt.Chart(...)
```

上面代码中，更新了轴上的 TitleFontSize 和 tiltleFont。同样，我们可以使用 alt.X( ) 和 alt.Y( ) 对象来自定义标题标签，如图 6.67 所示。

```
>>> (chart.configure_axis(titleFontSize=20,
...                       titleFont='Consolas')
...       .mark_point()
...       .encode(x=alt.X('Displacement',
...                       title='Displacement (m)'),
...               y=alt.Y('Acceleration',
...                       title='Acceleration (m/s)'),
...               color='Origin')
... )
alt.Chart(...)
```

注意，上面的代码中每个标签都有单位。如果我们想要控制垂直轴，则可以指定 configure_axisLeft，只有该轴会受到这些更改的影响。此外还有 configure_axisTop、configure_axisRight、configure_axisX、configure_axisY 等。轴上刻度标记的标签由 labelFontSize 和其他相关参数控制。

条形图的生成遵循相同的模式，只是使用了 mark_bar，如下所示（见图 6.68）：
```
>>> chart.mark_bar().encode(y='Origin',x='Horsepower')
alt.Chart(...)
```

同样，可以使用 mark_area 等生成面积图。你可以在调用结束时添加 interactive( )，使图标可通过使用鼠标滚轮进行缩放。图表可以自动保存为 PNG 和 SVG 格式，但这需要额外的浏览器自动化工具或 altair_saver。图表也可以通过必要的 JavaScript 嵌入保存到 HTML 文件中。

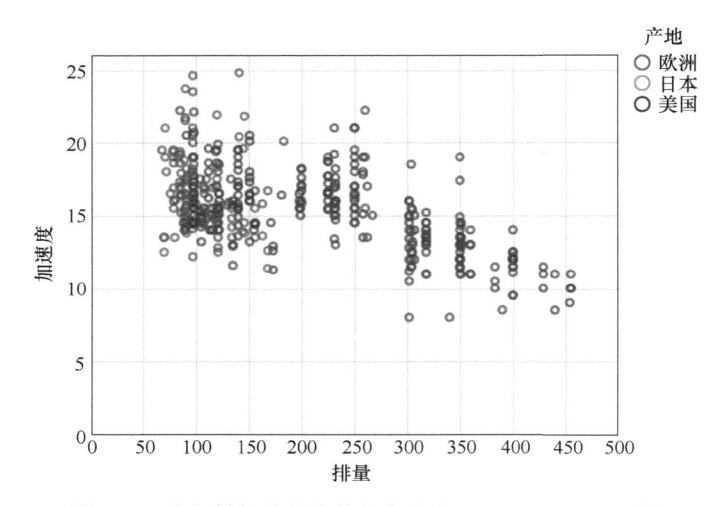

图 6.66　坐标轴标签的字体和大小由 configure_axis 设置

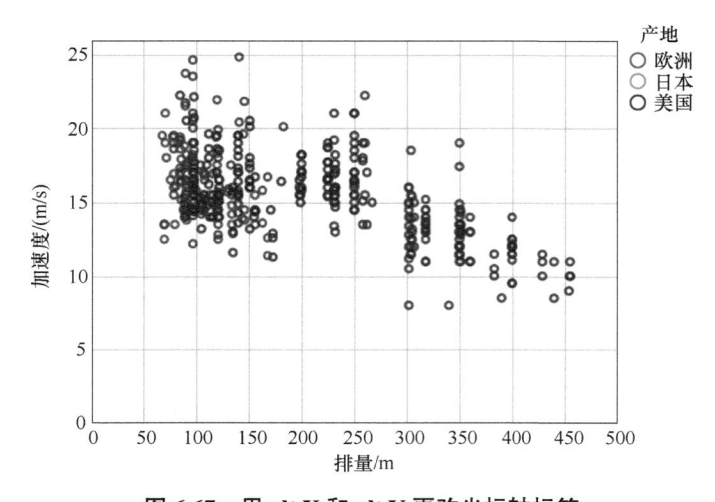

图 6.67　用 alt.X 和 alt.Y 更改坐标轴标签

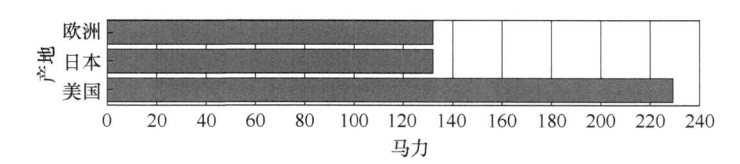

图 6.68　通过在数据框中指定的 x 和 y 列并使用 mark_bar 生成的条形图

6.4.2　聚合和转换

Altair 可以对 Dataframe 中的数据元素执行某些聚合操作。这样可以避免将聚合添加到另一个需要作为输入传递的单独数据框中（见图 6.69）。

```
>>> cars = vega_datasets.data('cars')
>>> (alt.Chart(cars).mark_point(size=200,
...                            filled=True)
...         .encode(x='Year:T',
...                 y='mean(Horsepower)')
... )
alt.Chart(...)
```

字符串中的 mean 表示将使用 Horsepower 的平均值作为 y 值（见图 6.69）。后缀：T 表示数据框中的 Year 列应被视为时间戳。mark_point 函数的参数控制点标记的大小和填充属性。

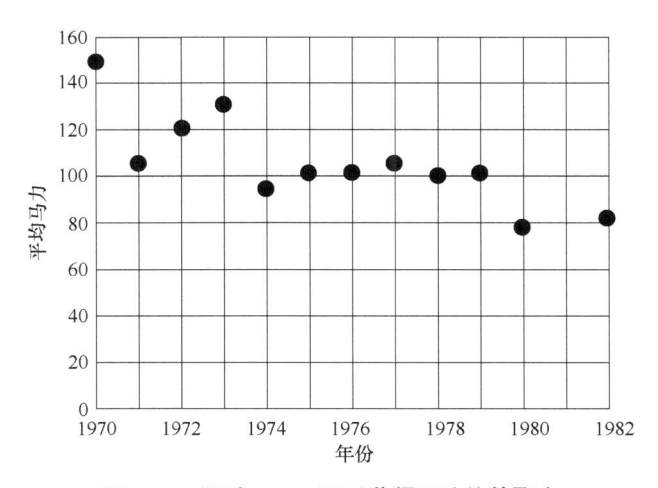

图 6.69　通过 Altair 可以获得平均值等聚合

在图 6.69 中，我们计算了 Horsepower 的平均值，其包括了所有来源的车辆。如果我们希望只包括美国制造的车辆，那么可以使用 transform_filter 方法，如下所示（见图 6.70）：

```
>>> (alt.Chart(cars).mark_point(size=200,filled=True)
...        .encode(x='Year:T',
...                y='mean(Horsepower)',
...                )
...        .transform_filter('datum.Origin=="USA"')
... )
alt.Chart(...)
```

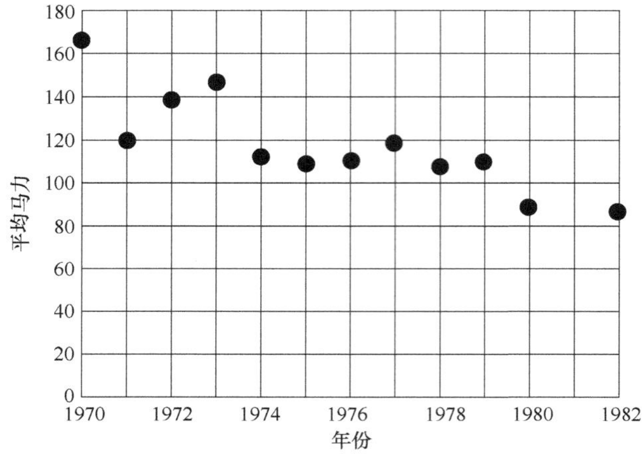

图 6.70 使用 transform_filter 的聚合

上面的代码中，transform_filter 确保只有美国车辆包括在平均值计算中。单词 datum 是 Vega-Lite 引用其数据元素的方式。除了表达式外，Altair 还提供了可以执行高级过滤操作的对象。例如，使用 FieldRangePredicate，我们可以选择连续变量中的一系列值，如下所示（见图 6.71）：

```
>>> (alt.Chart(cars).mark_point(size=200,filled=True)
...     .encode(x='Year:T',
...             y='mean(Horsepower)',
...             )
...     .transform_filter(alt.FieldRangePredicate('Horsepower',
...                                                [75,100]))
... )
alt.Chart(...)
```

这意味着只有在 75~100 之间的 Horsepower 值在计算中被选择。使用 alt.Scale 和 domain 关键字，可以调整生成图表的比例，如下所示（见图 6.72）：

```
>>> (alt.Chart(cars).mark_point(size=200,
...                             filled=True)
...     .encode(x='Year:T',
...             y=alt.Y('mean(Horsepower)',
...                     scale=alt.Scale(domain=[60,110])),
...             )
...     .transform_filter(alt.FieldRangePredicate('Horsepower',
...                                                [75,100]))
...     .properties(width=300,height=200)
... )
alt.Chart(...)
```

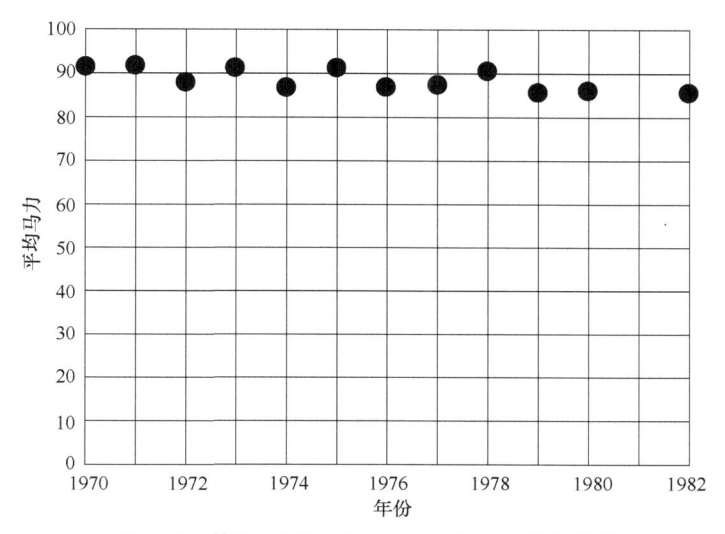

图 6.71　使用 **alt.FieldRangePredicate** 限定范围

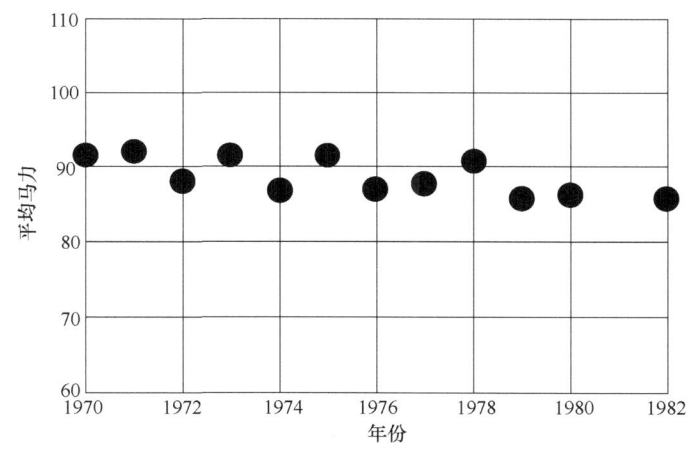

图 6.72　使用 **alt.Scale** 调整坐标轴范围

注意，生成的图形尺度限制已经使用 alt.Scale 和 domain 关键字进行了调整。谓词可以通过逻辑操作（如 LogicalNotPredicate 等）进行组合。

transform_calculate 方法允许将基本表达式应用于数据元素。例如，要计算马力的二次方，如下所示（见图 6.73）：

```
>>> from altair import datum
>>> h1=(alt.Chart(cars).mark_point(size=200,
...                                filled=True)
...        .encode(x='Year:T',
...                y='sqh:Q')
...        .transform_calculate(sqh = datum.Horsepower**2)
... )
```

注意，必须使用：Q（表示定量数据）来指定生成计算结果的类型。

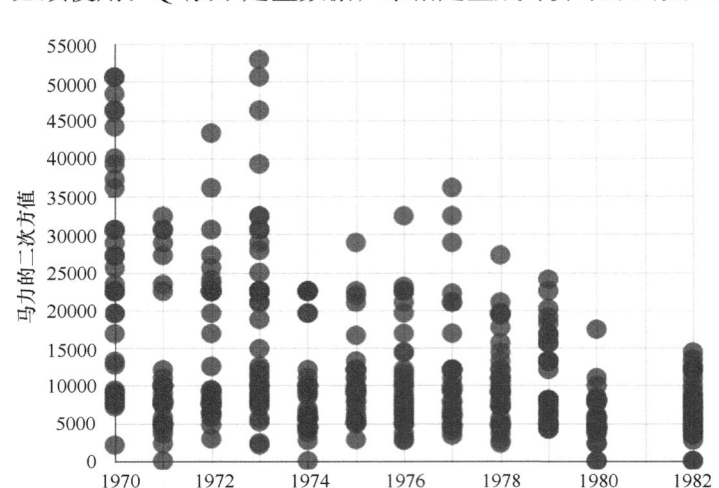

图 6.73 使用 transform_calculate 实现嵌入式计算

我们还可以使用 transform_aggregate 方法，利用前面 transform_calculate 方法得到的 sqh 变量，计算二次方的均值，同时按年份分组，如下所示（见图 6.74）：

```
>>> h2=(alt.Chart(cars).mark_point(size=200,
...                                filled=True,
...                                color='red')
...       .encode(x='Year:T',
...               y='msq:Q')
...       .transform_calculate(sqh = datum.Horsepower**2)
...       .transform_aggregate(msq='mean(sqh)',
...                            groupby=['Year'])
... )
```

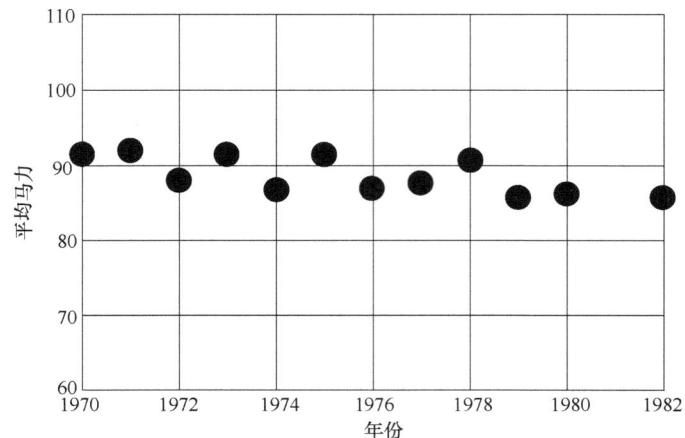

图 6.74 使用 transform_aggregate 用于计算中间项

两个图表可以使用加法运算符叠加在一起（见图 6.75）。
```
>>> h1+h2
alt.LayerChart(...)
```
transform_* 系列中还有其他函数，包括 transform_lookup、transform_window 和 transform_bin。主文档网站中有详细信息。

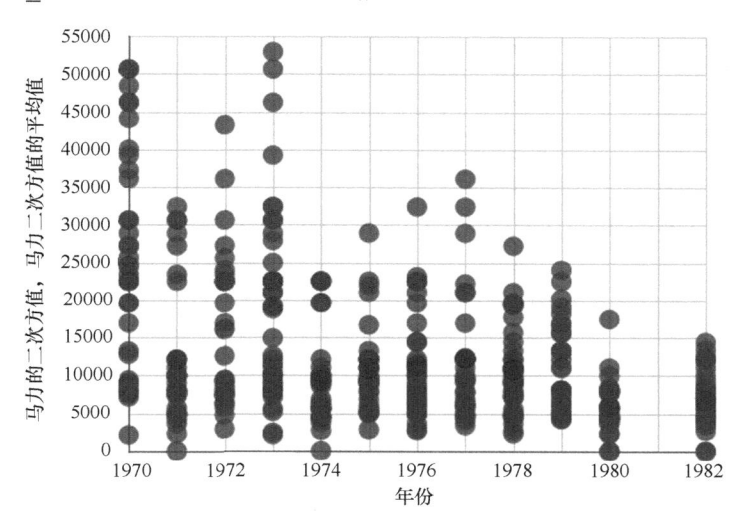

图 6.75　可以用加法运算符叠加图表

6.4.3　Altair 交互

Altair 具有从 Vega-Lite 继承的交互功能。这些功能通过工具的形式表达，可以轻松地分配给 Altair 可视化。交互的核心组件是 selection 对象。例如，以下代码创建一个 selection_interval( ) 对象（见图 6.76）：
```
>>> brush = alt.selection_interval()
```
现在，可以将刷选工具（brush）附加到 Altair 图表上（见图 6.76）。
```
>>> chart.mark_point().encode(y='Displacement',
...                           x='Horsepower')\
...      .properties(selection=brush)
alt.Chart(...)
```
在 Web 浏览器中，你可以在图表中拖动鼠标，会看到一个矩形选择框出现，但不会发生任何变化，因为我们还没有将 brush 与图表上的 action 关联起来。如果要更改图表上所选择对象的颜色，可以在 encode 方法中使用 alt.condition 方法（见图 6.77）。

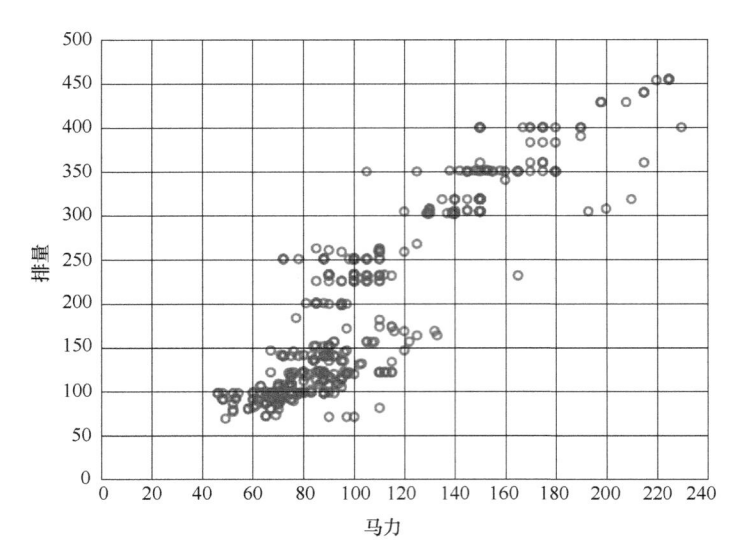

图 6.76　使用 alt.selection_interval 进行交互

```
>>> (chart.mark_point().encode(y='Displacement',
...                            x='Horsepower',
...                            color=alt.condition(brush,
...                                                'Origin:N',
...                                                alt.value('lightgray')))
...     .properties(selection=brush)
... )
alt.Chart(...)
```

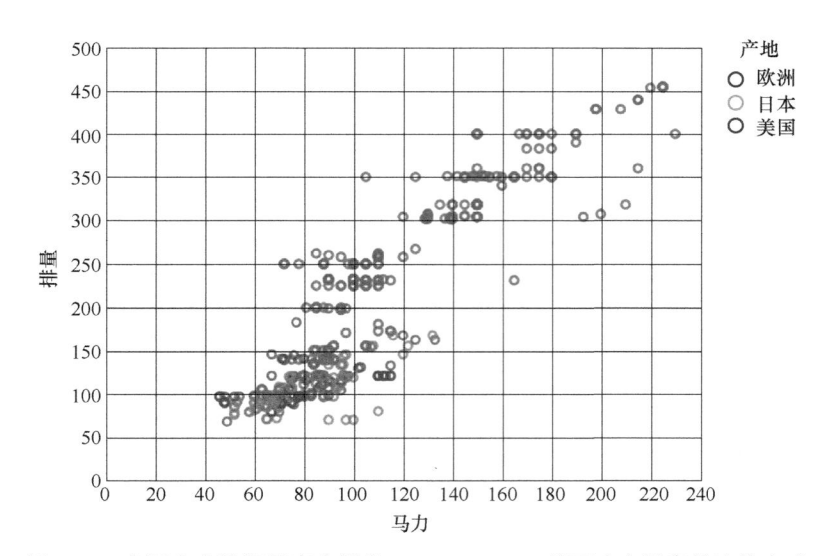

图 6.77　在图表中选择元素会触发 alt.condition，从而改变元素的渲染方式

上面的代码中，condition 意味着如果 brush selection 为 True，则根据它们的 Origin 值着色，否则使用 alt.value 设置为 lightgray，该值用于编码的设定。

选择器可以跨图表共享，如下所示（见图 6.78）：

```
>>> chart = (alt.Chart(cars)
...          .mark_point()
...          .encode(x='Horsepower',
...                  color=alt.condition(brush,
...                                      'Origin:O',
...                                      alt.value('lightgray')))
...          .properties(selection=brush)
...          )
>>> chart.encode(y='Displacement') & chart.encode(y='Acceleration')
alt.VConcatChart(...)
```

在上面的代码中，图表变量只定义了 x 坐标，并且嵌入了 cars 数据。& 符号将两个图表垂直排列。只有在最后一行，才为每个图表选择了 y 坐标。选择器在两个图表中都被共享，因此在其中一个图表中使用鼠标选择将导致选择（通过 alt.condition）在两个图表上高亮显示。这是一种复杂的交互方式，只需要很少的代码就可以实现。

其他的 selection 对象，如 alt.selection_multi 和 alt.selection_single，允许使用鼠标单击或鼠标悬停操作进行单个项的选择（例如，alt.selection_single（on='mouseover'））。对于 alt.selection_multi，需要按住 shift 键并单击鼠标进行选择。

Altair 是对快速变化的可视化领域的一种新颖而智能的处理方式。通过依托于 Vega-Lite，模块确保了能够跟上这一重要工作成果的步伐。Altair 引入了许多不属于标准 Matplotlib 或 Bokeh 词汇的可视化类型到 Python 中。然而，Altair 的

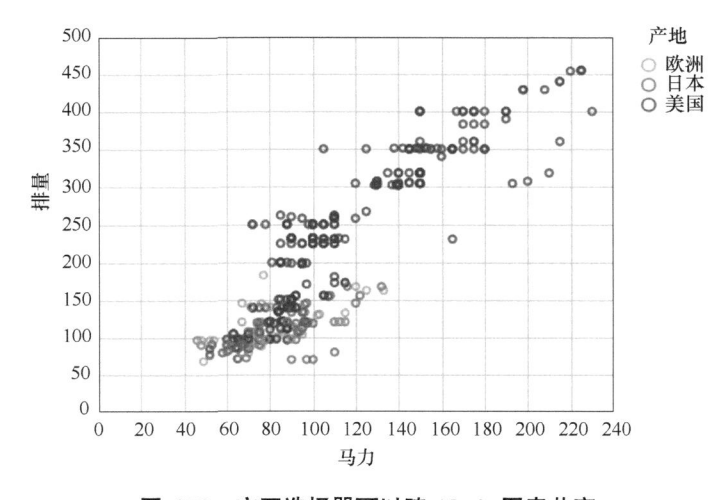

图 6.78 交互选择器可以跨 Altair 图表共享

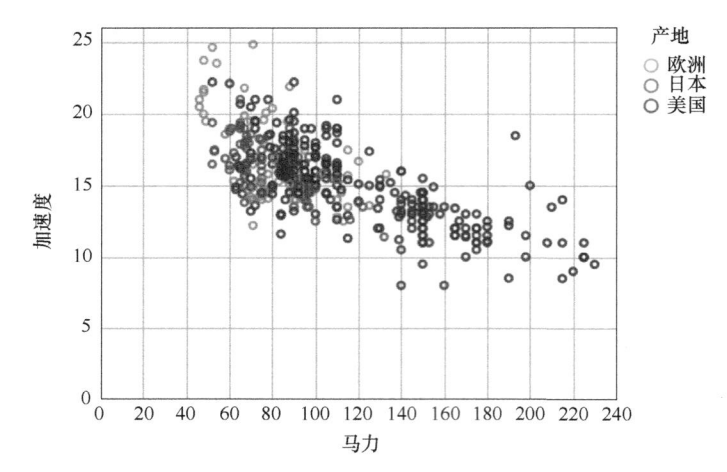

图 6.78　交互选择器可以跨 **Altair** 图表共享（续）

成熟度远低于 Matplotlib 或 Bokeh。Vega-Lite 本身是一个复杂的包，具有自己的 JavaScript 依赖。当出现问题时很难修复，因为问题可能过于深奥且分布在 JavaScript 堆栈中，Python 程序员可能无法触及（甚至无法识别出问题所在）。

6.5　Holoviews

Holoviews 提供了一种注释数据的方式，以便在下游使用 Bokeh、Plotly 或 Matplotlib 进行可视化时更方便。其关键概念是为数据元素提供语义，从而支持数据可视化的构建。这是一种类似于 Altair 的声明性方法，但与 Altair 不同的是，Holoviews 对后续的可视化构建没有固定的目标，而 Altair 则固定于 Vega-Lite 作为下游目标。

```
>>> import numpy as np
>>> import pandas as pd
>>> import holoviews as hv
>>> from holoviews import opts
>>> hv.extension('bokeh')
```

hv.extension 声明 Bokeh 应该被用作下游的可视化构造函数。让我们创建一些数据。

```
>>> xs = np.linspace(0,1,10)
>>> ys = 1+xs**2
>>> df = pd.DataFrame(dict(x=xs, y=ys))
```

为了用 Holoviews 可视化这个 Dataframe，我们必须决定要渲染什么以及如何渲染它。一种方法是使用 Holoviews Curve (见图 6.79)。

```
>>> c=hv.Curve(df,'x','y')
>>> c
:Curve   [x]   (y)
```

Holoviews 的 Curve 对象声明数据来自将 x 映射到 y 列的连续函数。为了呈现这个图，我们使用内置的 Jupyter 显示功能。

```
>>> c
:Curve   [x]   (y)
```

注意，Bokeh 小组件已经包含在图表中。Curve 对象仍然保留着嵌入的 Dataframe，可以通过数据属性进行操作。注意，我们也可以使用 NumPy 数组、Python 列表或 Python 字典来代替 Pandas Dataframe。

```
>>> c.data.head()
      x     y
0  0.00  1.00
1  0.11  1.01
2  0.22  1.05
3  0.33  1.11
4  0.44  1.20
```

如果我们不想填充数据点之间的线段，则可以选择 Holoviews Scatter 而不是 Curve (见图 6.80)。

```
>>> s = hv.Scatter(df,'x','y')
>>> s
:Scatter   [x]   (y)
```

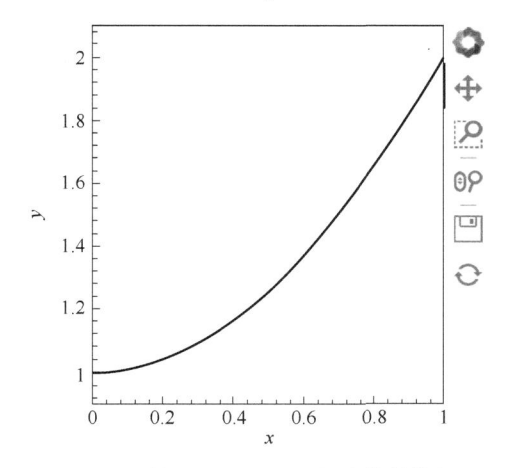

图 6.79　使用 Holoviews 曲线绘制数据　　图 6.80　使用 Holoviews Scatter 绘图

我们可以使用 + 运算符将两个图并排排列 (见图 6.81)。

```
>>> s+c  # 注意 Bokeh 工具是如何在绘图之间共享并影响两者的
:Layout
   .Scatter.I  :Scatter  [x]   (y)
   .Curve.I    :Curve    [x]   (y)
```

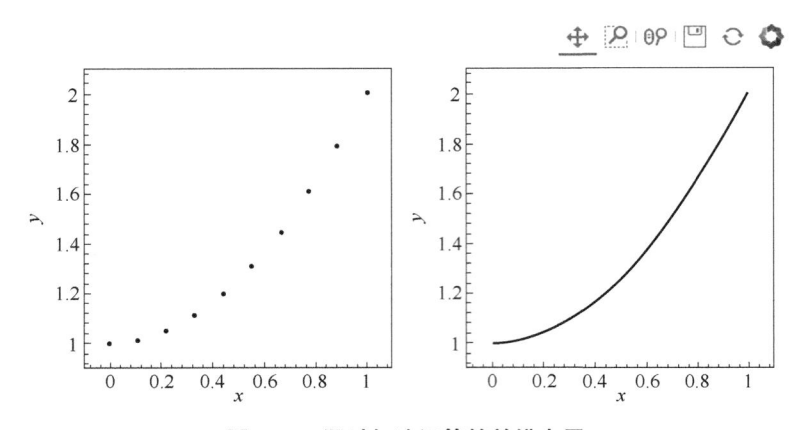

图 6.81 通过加法运算符并排布局

还可以使用乘法运算符覆盖（见图 6.82）。

```
>>> s*c # 乘法运算符的覆盖
:Overlay
   .Scatter.I  :Scatter   [x]    (y)
   .Curve.I    :Curve     [x]    (y)
```

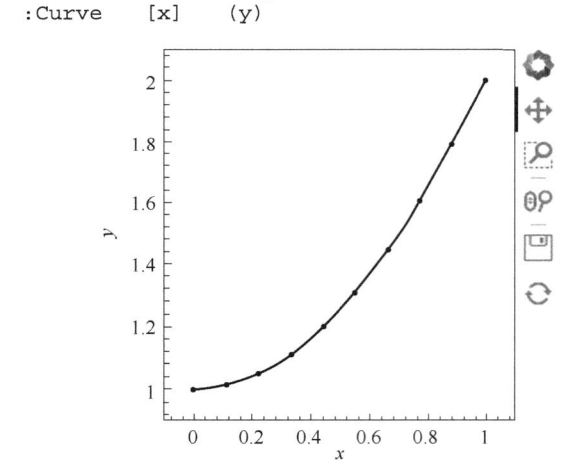

图 6.82 通过乘法运算符的覆盖图

图的宽度、高度等选项可以直接设定（见图 6.83）。

```
>>> s.opts(color='r',size=10)
:Scatter   [x]    (y)
>>> s.options(width=400,height=200)
:Scatter   [x]    (y)
```

你还可以在 Jupyter Notebook 中使用 %opts 命令来管理这些选项。虽然这是一种非常有条理的方法，但它仅在 Jupyter Notebook 内有效。如果要重设标题，则可以使用 relabel 方法（见图 6.84）。

```
>>> k = s.redim.range(y=(0,3)).relabel('Ploxxxt Title')  # 添加标题
>>> k.options(color='g',size=10,width=400,
...           height=200,yrotation=88)
:Scatter   [x]   (y)
```

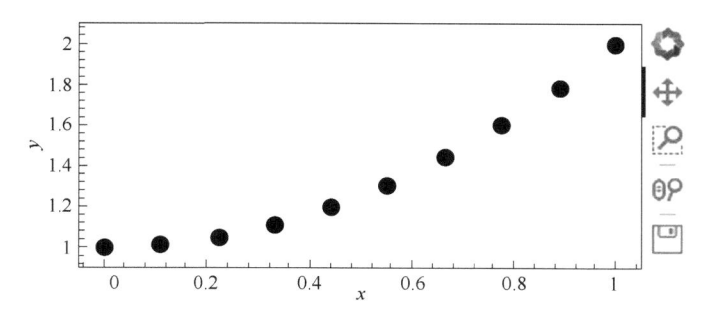

图 6.83 可视化尺寸可以在 options 方法中使用 width 和 height 设置

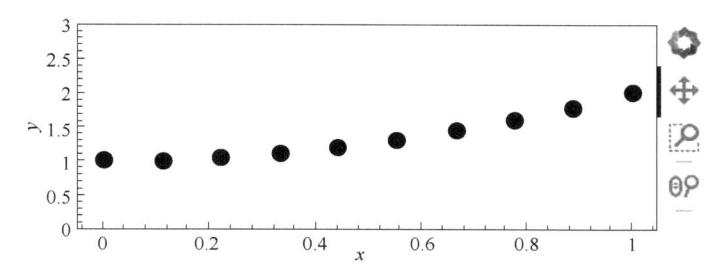

图 6.84 在 Holoviews 对象上使用 relabel 设置可视化标题

还有许多其他选项用于绘图，下面是使用 Bars 的示例（见图 6.85）：

```
>>> options = (opts.Scatter(width=400,
...                         height=300,
...                         xrotation=70,
...                         color='g',
...                         size=10),
...            opts.Bars(width=400,
...                      height=200,
...                      color='blue',
...                      alpha=0.3)
...           )
>>> (s.to(hv.Bars)*s).redim.range(y=(0,3)).options(*options)
:Overlay
   .Bars.I    :Bars    [x]    (y)
   .Scatter.I :Scatter [x]    (y)
```

上面的代码使用了 to 方法从 Scatter 对象创建了 Bars 对象。xrotation 改变了 x 轴标签的方向。

Holoviews 对象还可以进行数据切片，如下所示（见图 6.86）：

```
>>> s[0:0.5] + s.select(x=(0.5,None)).options(color='b')
:Layout
   .Scatter.I  :Scatter [x]    (y)
   .Scatter.II :Scatter [x]    (y)
```

上面的代码中，索引是基于 x 坐标的，None 表示直到 x 数据的末尾。

Holoviews 中将 key 维度作为 kdims，value 维度作为 vdims。这些维度用于分类哪些元素应被视为独立变量（kdims）或依赖变量（vdims）在图表中。

```
>>> s.kdims,s.vdims
([Dimension('x')], [Dimension('y')])
```

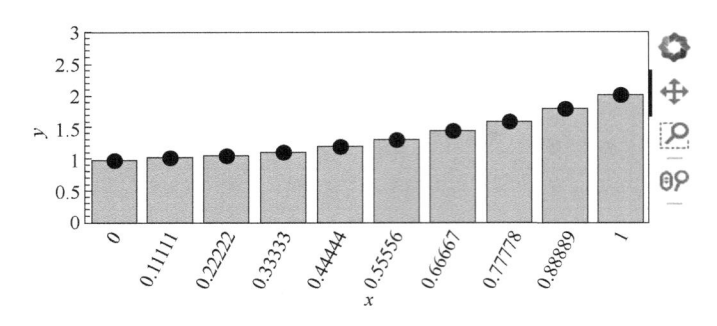

图 6.85　条形图在 Holoviews 中可用

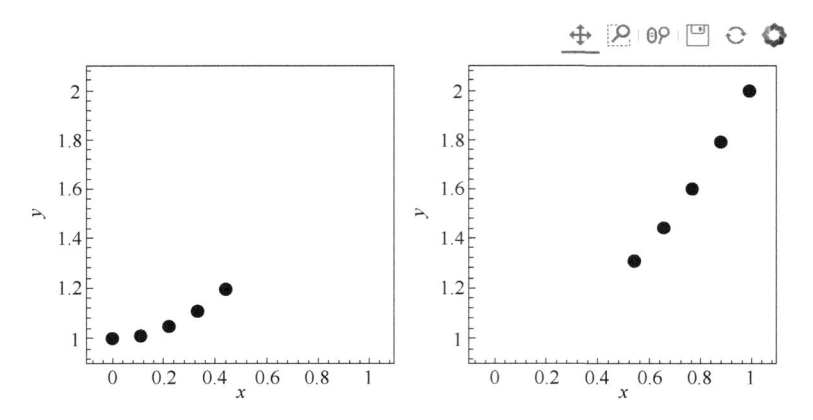

图 6.86　Holoviews 图中的数据可以切片并自动渲染

6.5.1　数据集

Holoviews 的 Dataset 并不是内置的元数据来直接支持可视化。Holoviews 的 Dataset 是一种创建一组维度的方式，这些维度将在稍后在下游可视化中被继承。在下面的例子中，我们只需要指定关键维度，其他列将被推断为值维度（vdims），可以稍后使用 groupby 进行分组。

```
>>> economic_data = pd.read_csv('macro_economic_data.csv')
>>> edata = hv.Dataset(data=economic_data,
...                    kdims=['country','year'])
>>> edata.groupby('year')
:HoloMap   [year]
   :Dataset   [country]    (growth,unem,capmob,trade)
>>> edata.groupby('country')
:HoloMap   [country]
   :Dataset   [year]    (growth,unem,capmob,trade)
```

注意，groupby 输出显示独立变量（即关键维度）year 或 country 所对应的其他值维度。

```
>>> edata.to(hv.Curve,
...          'year',
...          'unem',
...          groupby='country').options(height=200)
:HoloMap   [country]
   :Curve   [year]    (unem)
```

上面的 to 方法为 country 创建了一个相应的下拉小组件（见图 6.87）。

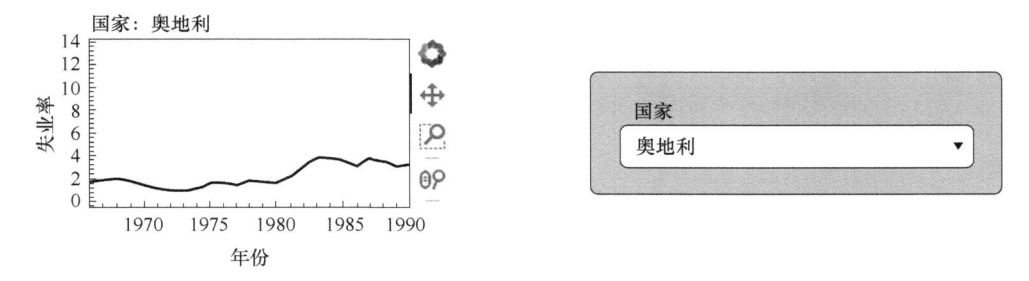

图 6.87　Holoviews 可以一步完成数据分组和绘图

我们还可以按照其他关键维度进行分组，例如创建一个 year 的滑块工具（见图 6.88）。注意，小组件的类型是根据关键维度的类型推断出来的（即，连续 year 的滑块和离散 country 的下拉窗口）。

```
>>> (edata.sort('country')
...       .to(hv.Bars,'country','unem',groupby='year')
...       .options(xrotation=45,height=300))
:HoloMap   [year]
   :Bars   [country]   (unem)
```

我们还可以不使用任何关键维度声明，以使它们都进入相应的小组件，如图 6.89 所示。

```
>>> edata.sort('country').to(hv.Bars,'growth','trade')
:HoloMap   [country,year]
   :Bars   [growth]   (trade)
```

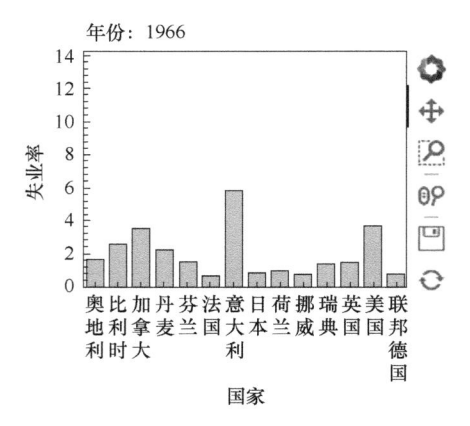

图 6.88 Holoviews 根据剩余维度中的数据类型自动选择滑块小组件

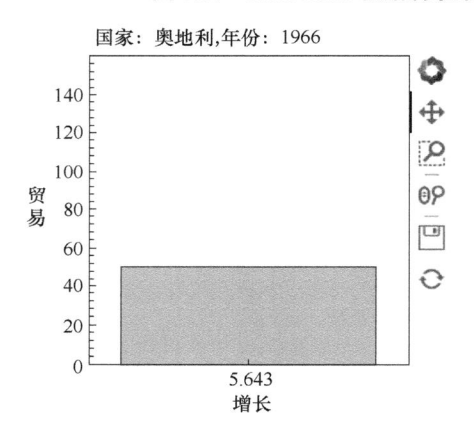

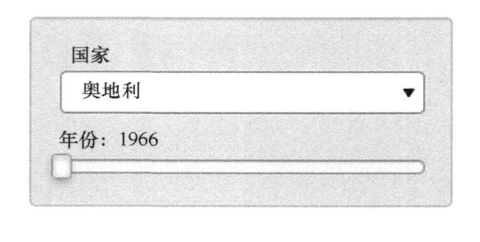

图 6.89 Holoviews 根据关键维度选择小组件

6.5.2 图像数据

Holoviews 元素可以处理二维表格或基于网格的数据，例如（见图 6.90）：

```
>>> x = np.linspace(0, 10, 500)
>>> y = np.linspace(0, 10, 500)
>>> z = np.sin(x[:,None]*2)*y
>>> image = hv.Image(z)
>>> image
:Image    [x,y]    (z)
```

我们可以使用 hist 方法获得图像颜色的直方图（见图 6.91）。

```
>>> image.hist()
:AdjointLayout
   :Image       [x,y]    (z)
   :Histogram   [z]      (z_count)
```

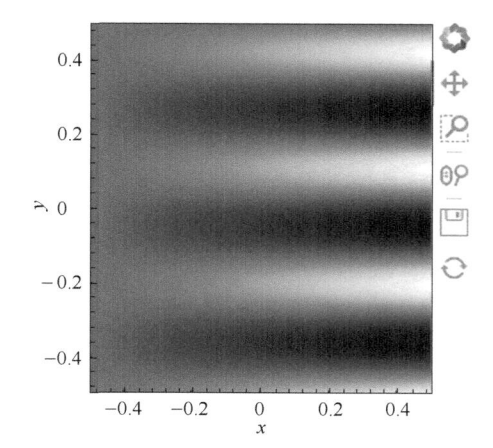

图 6.90　Holoviews 创建热图

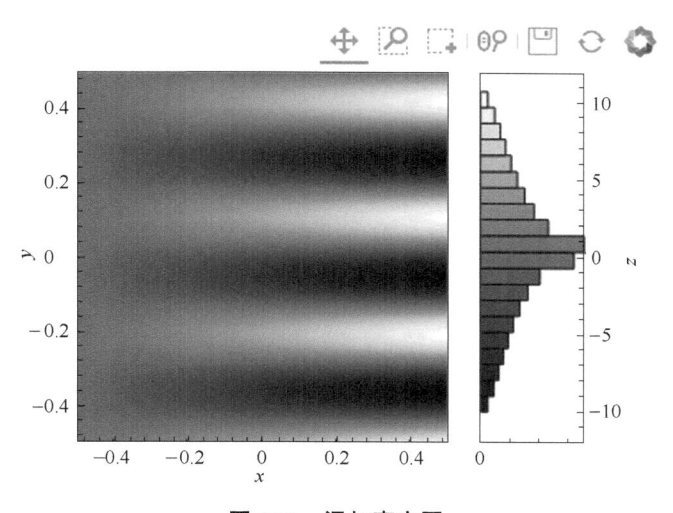

图 6.91　添加直方图

6.5.3　表格数据

```
>>> economic_data= pd.read_csv('macro_economic_data.csv')
>>> economic_data.head()
         country  year  growth  unem  capmob  trade
0  United States  1966    5.11  3.80       0   9.62
1  United States  1967    2.28  3.80       0   9.98
2  United States  1968    4.70  3.60       0  10.09
3  United States  1969    2.80  3.50       0  10.44
4  United States  1970   -0.20  4.90       0  10.50
```

元素分组可通过将 vdims 指定为列名列表，并为 Holoviews 对象提供 color_index 列名和 color（hv.Palette.colormaps.keys（））实现（见图 6.92）。注意，指定 hover 工具意味着在鼠标悬停在元素上时显示每个元素的数据。另外，除非图形足够高以容纳所有图例项，否则不会显示所有图例项。

```
>>> options = opts.Scatter(tools=['hover'],
...                        legend_position='left',
...                        color_index='country',
...                        width=800,height=500,
...                        alpha=0.5,
...                        color=hv.Palette('Category20'),
...                        size=10)
>>> c =
↪   hv.Scatter(economic_data,'year',['trade','country','unem'])
>>> c.redim.range(trade=(0,180)).options(options)
:Scatter   [year]   (trade,country,unem)
```

使用 Dataset 也适用于二维可视化，如热图。将鼠标悬停在图 6.93 上，显示每个单元格的相应数据。

```
>>> options = opts.HeatMap(colorbar=True,
...                        width=600,
...                        height=300,
...                        xrotation=60,
...                        tools=['hover'])
>>>
↪   edata.to(hv.HeatMap,['year','country'],'growth').options(options)
:HeatMap   [year,country]   (growth)
```

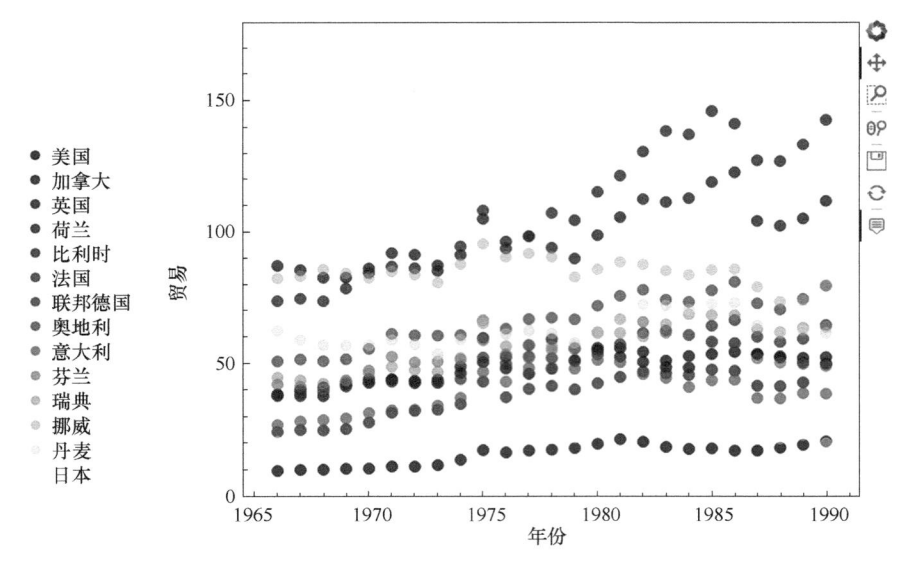

图 6.92　Holoviews014

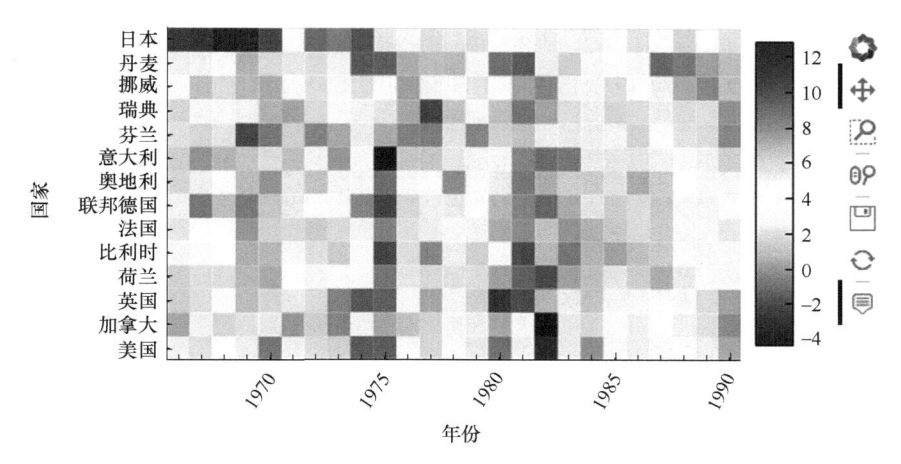

图 6.93 Holoviews Dataset 支持二维可视化

6.5.4 自定义交互

DynamicMap 为 Holoviews 可视化提供了交互性，角度键的维度由滑块控件提供。命名的函数是惰性求值的（见图 6.94）。

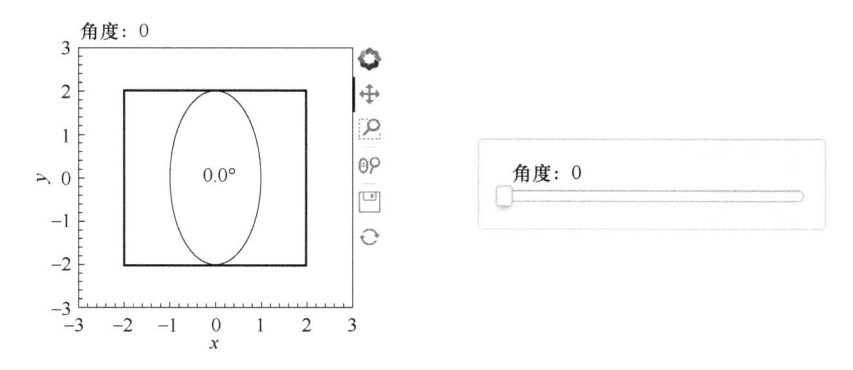

```
>>> def dynamic_rotation(angle):
...     radians = (angle / 180) * np.pi
...     return (hv.Box(0,0,4,orientation=-radians).
options(color='r',line_width=3)
...             * hv.Ellipse(0,0,(2,4), orientation=radians)
...             * hv.Text(0,0,'{0}°'.format(float(angle))))
...
>>> hv.DynamicMap(dynamic_rotation,
...               kdims=['angle']).redim.range(angle=(0, 360),
...                                            y=(-3,3),
...                                            x=(-3,3))
:DynamicMap   [angle]
```

图 6.94 Holoviews 为指定的关键维度构建小组件

图 6.95 是另一个使用相应滑块小组件的示例。注意，通过将滑块小组件变量范围声明为浮点数，可以获得函数滑块移动的更高分辨率。

```
>>> def sine_curve(f=1,phase=0,ampl=1):
...     xi = np.linspace(0,1,100)
...     y = np.sin(2*np.pi*f*xi+phase/180*np.pi)*ampl
...     return hv.Curve(dict(x=xi,y=y)).redim.range(y=(-5,5))
...
>>> hv.DynamicMap(sine_curve,
...               kdims=['f','phase','ampl'])\
...                 .redim.range(f=(1,3.),
...                              phase=(0,360),
...                              ampl=(1,3.))
:DynamicMap    [f,phase,ampl]
```

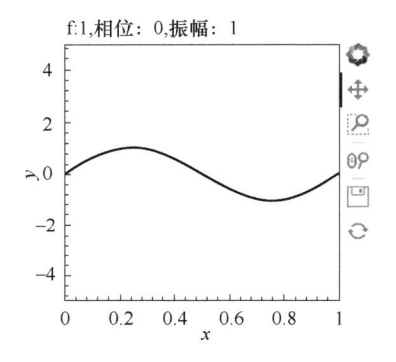

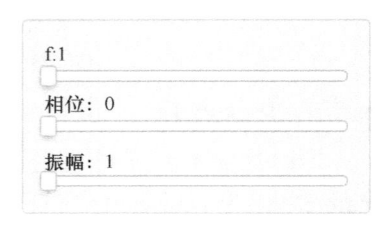

图 6.95 自动创建的 Holoviews 小组件是根据绘制维度的类型派生的

6.5.5 流

Holoviews 使用流来将数据传递给容器或元素，而不是使用滑块进行输入。一旦定义了流，则 hv.DynamicMap 可以进行绘制（见图 6.96）。

```
>>> from holoviews.streams import Stream
>>> F = Stream.define('Freq',f=3)
>>> Phase = Stream.define('phase',phase=90)
>>> Amplitude = Stream.define('amplitude ',ampl=1)
>>> dm=hv.DynamicMap(sine_curve,streams = [F(f=1),
...                                        Phase(phase=0),
...                                        Amplitude(ampl=1)])
>>> dm
:DynamicMap    []
```

流是通过发送事件来触发的。

```
>>> # 绘图更新
>>> dm.event(f=2,ampl=2,phase=30)
```

或者使用非阻塞的 g.periodic（）方法。

```
>>> dm.periodic(0.1,count=10,timeout=8,param_fn=lambda
↪ i:{'f':i,'phase':i*180})
```

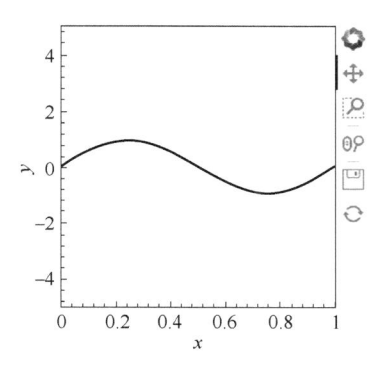

图 6.96　Holoviews DynamicMap 处理流

6.5.6　Pandas 与 hvplot 集成

Holoviews 可以与 Pandas Dataframe 集成，通过在 Pandas Dataframe 对象上放置额外的绘图选项来加速常见的绘图场景。

```
>>> import hvplot.pandas
>>> df.head()
      x     y
0  0.00  1.00
1  0.11  1.01
2  0.22  1.05
3  0.33  1.11
4  0.44  1.20
```

可以直接从 Dataframe hvplot() 方法生成图（见图 6.97）。

```
>>> df.hvplot()  # 现在使用的是 Bokeh 后端而不是 matplotlib
:NdOverlay   [Variable]
   :Curve   [index]    (value)
```

下面使用 Holoviews 绘制分组图。注意，图表上没有 y 标签。这是因为通过分组创建的中间对象是一个没有标记列的 Series 对象。水平条形图（barh()）如图 6.98 所示。

```
>>> economic_data.groupby('country')['trade'].sum().hvplot.barh()
:Bars   [country]   (trade)
```

注意中间输出的类型，

```
>>> type(economic_data.groupby('country')['trade'].sum())
<class 'pandas.core.series.Series'>
```

修复丢失的 y 标签的一种方法是使用 to_frame() 将中间的 Series 对象转换

为 Dataframe，为求和添加一个带标签的列（见图 6.99）。

```
>>> (economic_data.groupby('country')['trade'].sum()
...                                           .to_frame('trade
↪   (units)')
...                                           .hvplot.barh()
... )
:Bars   [country]   (trade (units))
```

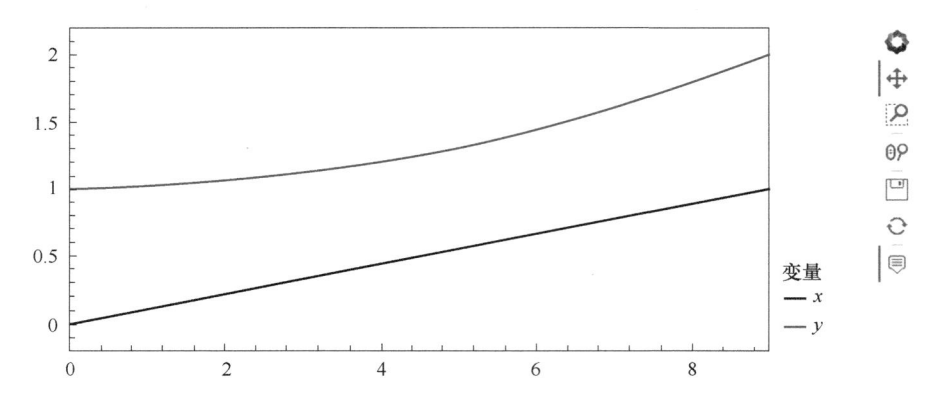

图 6.97　使用 hvplot 直接从 Pandas Dataframe 生成的 Holoviews 图

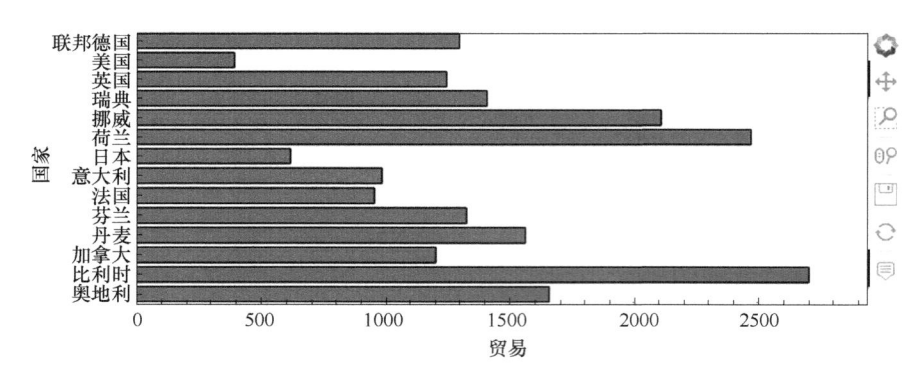

图 6.98　使用 groupby 的 Holoviews 条形图

我们还可以使用 Pandas 通过 sort_values( ) 来按值进行排序，如图 6.100 所示。

```
>>> (economic_data.groupby('country')['trade'].sum()
...     .sort_values()
...     .to_frame('trade(units)')
...     .hvplot.barh()
... )
:Bars   [country]   (trade(units))
```

也可以展开分组，如图 6.101 所示。

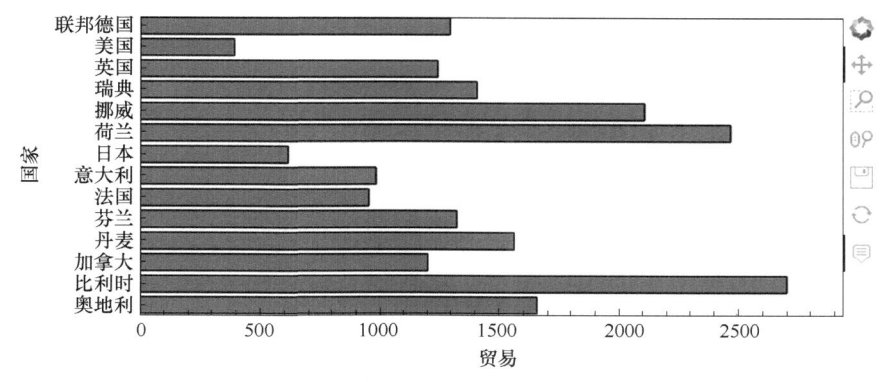

图 6.99 可视化中 Pandas Dataframe 标签的列名

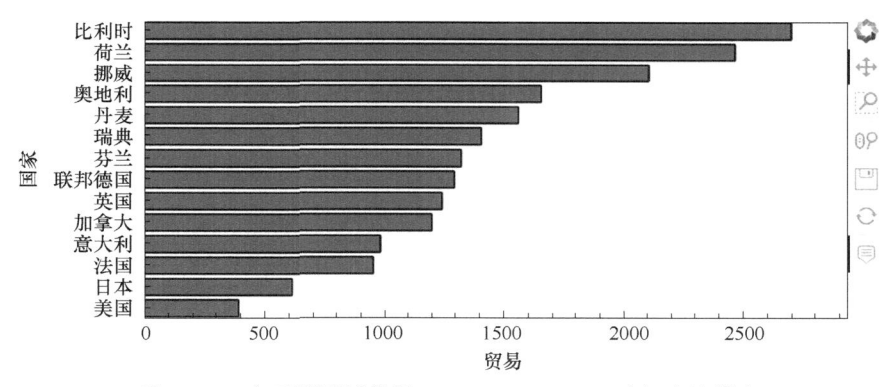

图 6.100 条形图可以使用 Pandas sort_values() 方法排序

```
>>> # 这是一个未堆叠的分组
>>> (economic_data.groupby(['year','country'])['trade']
...     .sum()
...     .unstack()
...     .head()
... )
country  Austria  Belgium  Canada  Denmark  ...  Sweden  United Kingdom  United States  West Germany
year                                        ...
1966       50.83    73.62   38.45    62.29  ...   44.79           37.93           9.62         37.89
1967       51.54    74.54   40.16    58.78  ...   43.73           37.83           9.98         38.81
1968       50.88    73.59   41.07    56.87  ...   42.46           37.76          10.09         39.51
1969       51.63    78.45   42.77    56.77  ...   43.52           41.93          10.44         41.40
1970       55.52    84.38   44.17    57.01  ...   46.28           42.80          10.50         43.07

[5 rows x 14 columns]
```

我们还可以为条形图添加颜色，如图 6.102 所示。注意，上面的代码中使用了 hv.Dimension 来修正 y 标签的缩放，用 rot 关键字参数更改了刻度标签的方向。

```
>>> (economic_data.groupby(['year','country'])['trade']
...     .sum()
...     .unstack()
...     .hvplot.bar(stacked=True,rot=45)
...     .redim(value=hv.Dimension('value',label='trade',
range=(0,1000)))...  )
:Bars   [year,Variable]   (value)
```

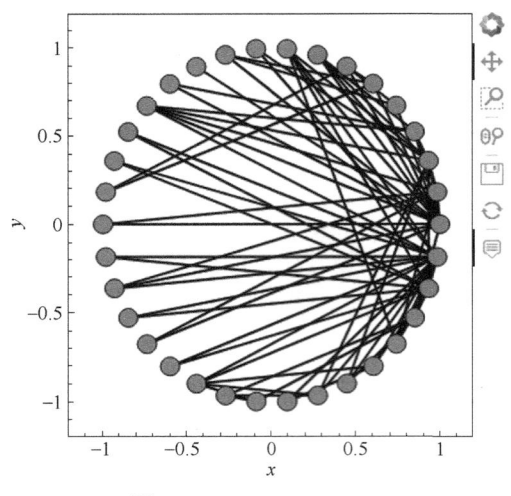

图 6.101　Holoviews025

在图 6.102 中，我们可以通过使用 variable 作为 color_index 来修复这个问题。因为 dataframe 没有为单元格中的值提供对应的名称，y 轴标签仍然无法访问，但通过使用 relabel，至少我们可以在 y 轴附近添加一个标题（见图 6.103）。

```
>>> options = opts.Bars(tools=['hover'],
...                     legend_position='left',
...                     color_index='Variable',
...                     width=900,
...                     height=400)

>>> (economic_data.groupby(['year','country'])['trade']
...  .sum()
...  .unstack()
...  .hvplot.bar(stacked=True,rot=45)
...  .redim(value=hv.Dimension('value',label='trade',range=(0,1000)))
...  .relabel('trade(units)').options(options)
... )
:Bars   [year,Variable]   (value)
```

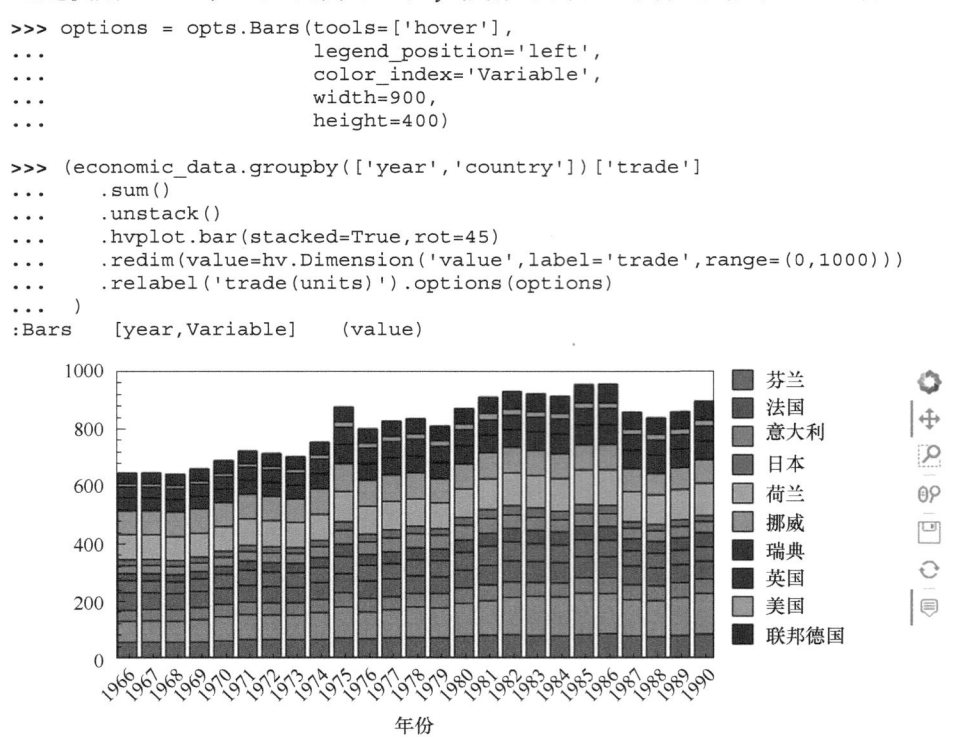

图 6.102　hv.Dimension 实现对坐标和标签的设置

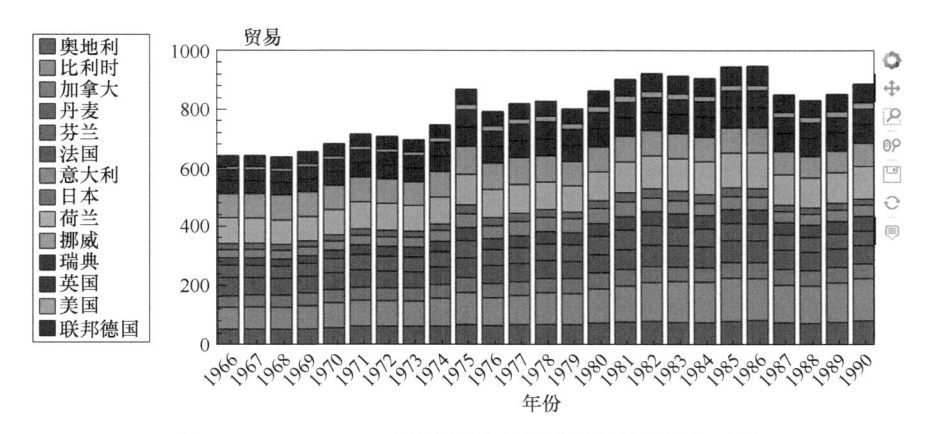

图 6.103 Holoviews 可以用很少的代码绘制堆叠条形图

Holoviews 还可以通过切片方式来对条形图的数据显示范围进行选择，例如 x 轴仅显示 1980 年以后的年份（见图 6.104）。

```
>>> options = opts.Bars(tools=['hover'],
...                     legend_position='left',
...                     color_index='Variable',
...                     width=900,
...                     height=400)
>>> (economic_data.groupby(['year','country'])['trade']
...    .sum()
...    .unstack()
...    .hvplot.bar(stacked=True,rot=45)
...    .redim(value=hv.Dimension('value',
...                              label='trade',
...                              range=(0,1000)))
...    .relabel('trade(units)').options(options)[1980:]
... )
:Bars    [year,Variable]    (value)
```

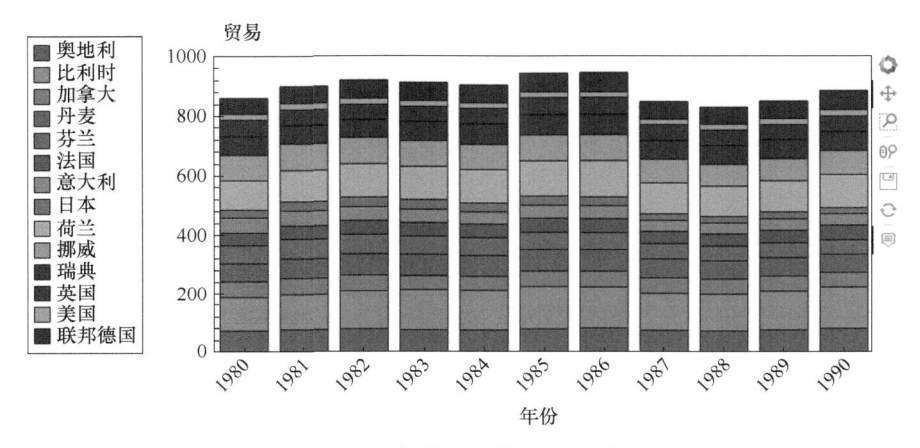

图 6.104 Holoviews 可视化可以像 Numpy 数组一样被切片

使用 hvplot 标签的主要问题在于 groupby 中的 reduce 操作没有 Holoviews 可以抓取的名称，这使得应用标签变得困难。下面的代码示例中，数据被重新调整为更适合 Holoviews 的格式（见图 6.105）：

```
>>> options = opts.Bars(tools=['hover'],
...                     legend_position='left',
...                     color_index='country',
...                     width=900,
...                     stacked=True,
...                     fontsize=dict(title=18,
...                                   ylabel=16,
...                                   xlabel=16),
...                     height=400)
>>> k=(economic_data.groupby(['year','country'])['trade']
...      .sum().
...      to_frame().T
...      .melt(value_name='trade'))
>>> (hv.Bars(k,kdims=['year','country'])
...      .options(options)
...      .relabel('Trade').redim.range(trade=(0,1200)))
:Bars    [year,country]   (trade)
```

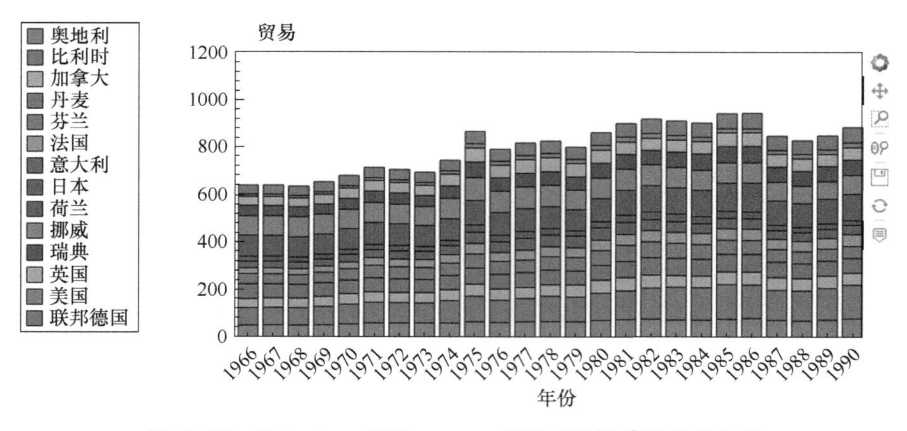

图 6.105　Holoviews 使用 Pandas 列的名称和索引作为标签

6.5.7　网络图

Holoviews 提供了制作网络图的工具，并与 networkx 很好地集成在一起。基本代码设置如下：

```
>>> import networkx as nx
>>> defaults = dict(width=400, height=400, padding=0.1)
```

```
>>> hv.opts.defaults(opts.EdgePaths(**defaults),
...                  opts.Graph(**defaults),
...                  opts.Nodes(**defaults))
```

我们可以使用 networkx 创建图 6.106。注意，通过将鼠标悬停在圆上，可以获得嵌入图中的节点的信息。另外，悬停还会显示节点的索引。

```
>>> G = nx.karate_club_graph()
>>> hv.Graph.from_networkx(G,
...
↪   nx.layout.circular_layout).opts(tools=['hover'])
:Graph    [start,end]
```

为了进一步理解 networkx，我们创建一个独立的图表。注意，nodes 和 edgepaths 是 Holoviews 内部表示图形的方式。nodes 以方括号显示图中的自变量，以括号显示因变量。在这个例子中，club 是图的节点属性。edgepaths 是由 nx.layout.circular_layout 确定的单个节点的位置。

```
>>> H=hv.Graph.from_networkx(G, nx.layout.circular_layout)
>>> H.nodes
:Nodes    [x,y,index]    (club)
>>> H.edgepaths
:EdgePaths    [x,y]
```

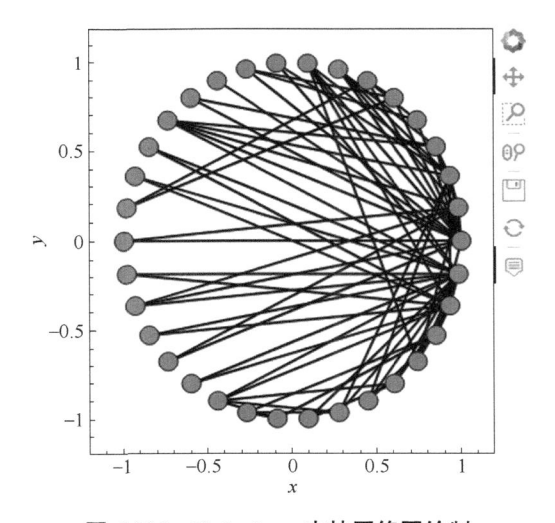

图 6.106　Holoviews 支持网络图绘制

让我们向 networkx 图添加一些边的权重，然后重建 Holoviews 图（见图 6.107）。

```
>>> for i,j,d in G.edges(data=True):
...     d['weight'] = np.random.randn()**2+1
...
>>> H=hv.Graph.from_networkx(G, nx.layout.circular_layout)
>>> H
:Graph    [start,end]    (weight)
```

注意，现在将鼠标悬停在边上会显示边权重，但你不能再像在以前的渲染中那样检查节点。

```
>>> H.opts(inspection_policy='edges')
:Graph   [start,end]   (weight)
```

注意，边缘在鼠标悬停时会高亮显示。

```
>>> H.opts(inspection_policy='nodes',
...         edge_color_index='weight',
...         edge_cmap='hot')
:Graph   [start,end]   (weight)
```

我们还可以根据权重对边着色。

```
>>> H.opts(inspection_policy='nodes',
...         edge_color_index='weight',
...         edge_cmap='hot')
:Graph   [start,end]   (weight)
```

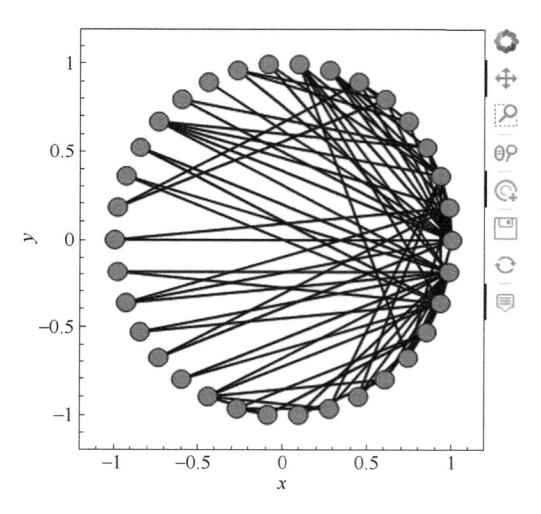

图 6.107　Holoviews 网络图添加边缘权重

我们还可以使用乘法运算符来叠加两个图形，并使用不同的策略来获取节点和边的悬停信息（见图 6.108）。

```
>>> H.opts(inspection_policy='edges', clone=True) * H
:Overlay
   .Graph.I   :Graph   [start,end]   (weight)
   .Graph.II  :Graph   [start,end]   (weight)
```

还可以用 hv.dim 对边加粗（见图 6.109）。

```
>>> H.opts(edge_line_width=hv.dim('weight'))
:Graph   [start,end]   (weight)
```

可以使用 Set1 颜色表根据图中的 club 值为节点着色（见图 6.110）。

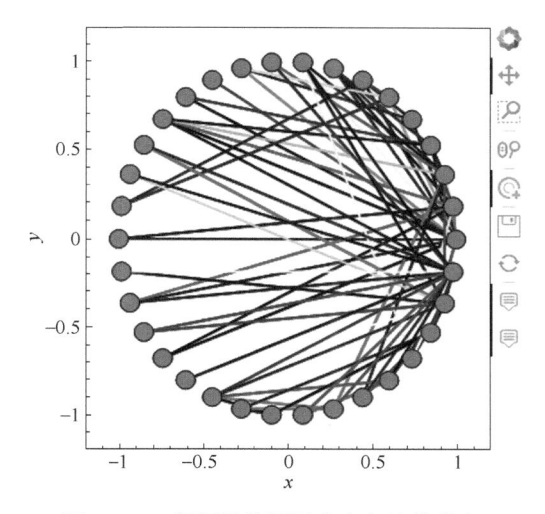

图 6.108 鼠标悬停显示节点与边缘信息

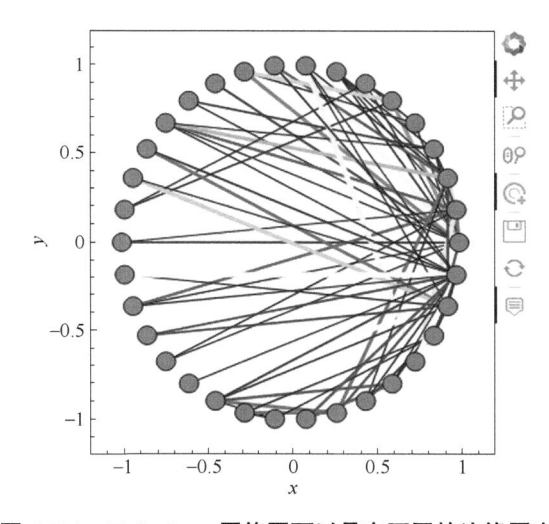

图 6.109 Holoviews 网络图可以具有不同的边缘厚度

```
>>> H.opts(node_color=hv.dim('club'),cmap='Set1')
:Graph   [start,end]   (weight)
```

生成最小生成树图，如图 6.111 所示。

```
>>> t = nx.minimum_spanning_tree(G)
>>> T=hv.Graph.from_networkx(t, nx.layout.kamada_kawai_layout)
>>> T.opts(node_color=hv.dim('club'),cmap='Set1',
...        edge_line_width=hv.dim('weight')*3,
...        inspection_policy='edges',
...        edge_color_index='weight',edge_cmap='cool')
:Graph   [start,end]   (weight)
```

保存 Holoviz 对象

Holoviews 提供了一个 hv.save（）方法，用于将渲染的图形保存到 HTML 文件中。只要没有依赖服务器的回调，生成的 HTML 就是一个静态文件，易于分发。

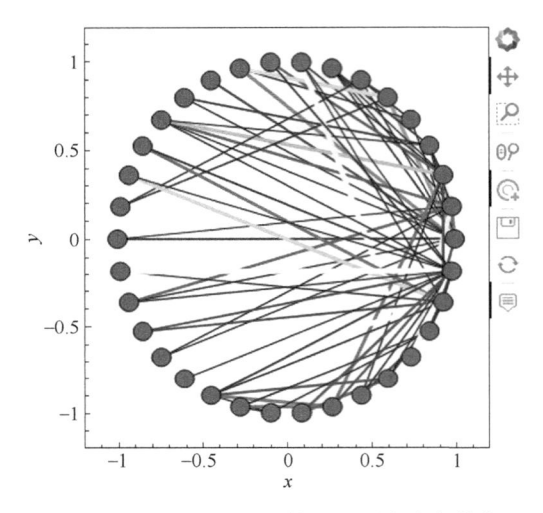

图 6.110　Holoviews 网络图可以有彩色节点

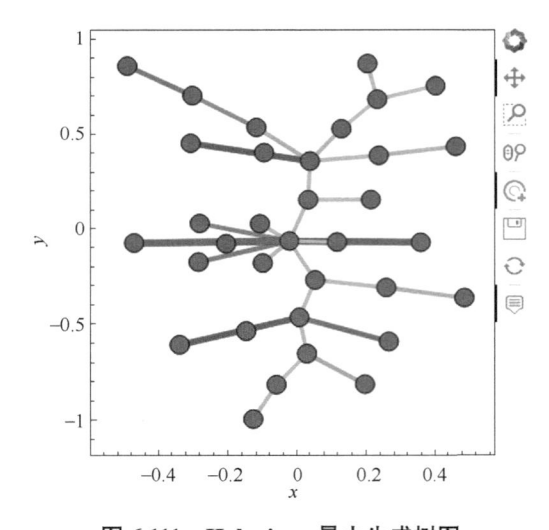

图 6.111　Holoviews 最小生成树图

6.5.8　Holoviz Panel

Panel 可以创建具有动态更新元素的仪表板。

```
>>> import panel as pn
>>> pn.extension()
```

与 ipywidgets 类似，Panel 具有用于将 Python 回调函数附加到小组件的 interact 功能。下面示例了如何根据滑块报告一个值（见图6.112）：

```
>>> def show_value(x):
...     return x
...
>>> app=pn.interact(show_value, x=(0, 10))
>>> app
Column
    [0] Column
        [0] IntSlider(end=10, name='x', value=5, value_throttled=5)
    [1] Row
        [0] Str(int, name='interactive07377')
```

pn.interact 返回的对象可以进行索引并重新定位。注意，文本出现在左侧，而不是底部（见图6.113）。

```
>>> print(app)
Column
    [0] Column
        [0] IntSlider(end=10, name='x', value=5, value_throttled=5)
    [1] Row
        [0] Str(int, name='interactive07377')
>>> pn.Row(app[1], app[0])  # 文本和小组件按行排列
Row
    [0] Row
        [0] Str(int, name='interactive07377')
    [1] Column
        [0] IntSlider(end=10, name='x', value=5, value_throttled=5)
```

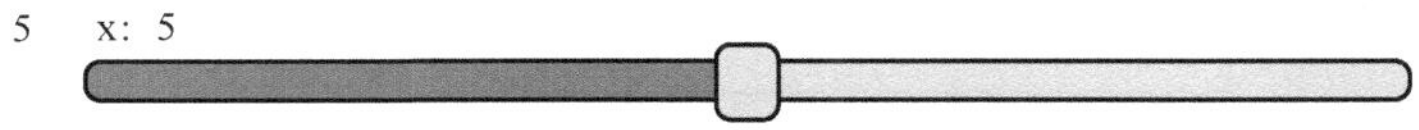

图 6.112　Panel interact 将回调函数连接到小组件

图 6.113　Holoviews Panel 支持文本标记

Panel 组件类型

Panel 中有三种主要的组件类型：

- Pane：Pane 包装外部对象（文本、图像、绘图等）的视图。
- Panel：Panel 在行、列或网格中布置多个组件。
- Widget：Widget 提供输入控件，向面板添加交互功能。

下面使用 pn.panel 对象对标记进行渲染（不包括 MathJaX）：

```
>>> pn.panel('### This is markdown **text**')
Markdown(str)
```

还可以将原始 HTML 作为面板元素。

```
>>> pn.pane.HTML('<marquee width=500><b>Breaking News</b>: some
↪   news.</marquee>')
HTML(str)
```

Pane 的主要布局机制是 pn.Row 和 pn.Column。还有 pn.Tabs 和 pn.GridSpec 用于更复杂的布局。

```
>>> pn.Column(pn.panel('### This is markdown **text**'),
...     pn.pane.HTML('<marquee width=500><b>Breaking News</b>:
↪   some news.</marquee>'))
Column
    [0] Markdown(str)
    [1] HTML(str)
```

我们可以使用小组件并连接到 pane。一种方法是使用 @pn.depends 装饰器，它将输入框的输入字符串连接到 title_text 回调函数（见图 6.114）。注意，必须按 ENTER 键才能更新文本。

```
>>> text_input = pn.widgets.TextInput(value='cap words')

>>> @pn.depends(text_input.param.value)
... def title_text(value):
...     return '## ' + value.upper()
...
>>> app2 = pn.Row(text_input, title_text)
>>> app2
```

图 6.114　**Panel decorator** 将回调连接到小组件

下面是一个自动补全小组件，注意，必须键入至少两个字符（见图 6.115 和图 6.116）。

```
>>> autocomplete = pn.widgets.AutocompleteInput(
...     name='Autocomplete Input',
...     options=economic_data.country.unique().tolist(),
...     placeholder='Write something here and <TAB> to complete')

>>> pn.Column(autocomplete,
...           pn.Spacer(height=50))  # 增加了一些垂直空间
```

```
Column
    [0] AutocompleteInput(name='Autocomplete
↪ Input',options=['United States', ...], placeholder='Write
↪ something h...)
    [1] Spacer(height=50)
```

自动补全输入

在这里输入内容并按<TAB>键进行补全

图 6.115 Panel 自动补全小组件

一旦你对应用程序满意，则可以使用 servable() 进行注释。在 Jupyter Notebook 中，此注释没有任何效果，但是使用 Jupyter Notebook（或纯 Python 文件）运行 panel serve 命令将创建一个带有注释仪表板的本地 Web 服务器。

通过使用这些元素，我们可以快速创建 economic_data 的仪表盘。注意 barchart 函数上的装饰器参数。你可以设置 IntRangeSlider 的极值，然后通过拖动所示小组件的中间来移动间隔。

```
>>> pulldown = (pn.widgets.Select(name='Country',
...                     options=economic_data.country
...                                         .unique()
...                                         .tolist()
...                )
>>> range_slider = pn.widgets.IntRangeSlider(start=1966,
...                                          end=1990,
...                                          step=1,
...                                          value=(1966,1990))

>>> @pn.depends(range_slider.param.value,pulldown.param.value)
... def barchart(interval, country):
...     start,end = interval
...     df=
↪ economic_data.query('country=="%s"'%(country))[['year','trade']]
...     return (df.set_index('year')
...             .loc[start:end]
...             .hvplot.bar(x='year',y='trade')
...             .relabel(f'Country: {country}'))
...
>>> app=pn.Column(pulldown,range_slider,barchart)
>>> app
Column
    [0] Select(name='Country', options=['United States', ...],
     value='United States')
    [1] IntRangeSlider(end=1990, start=1966, value=(1966, 1990),
     value_throttled=(1966, 1990))
    [2] ParamFunction(function)
```

注意，回调需要后端 Python 服务器来更新绘图。

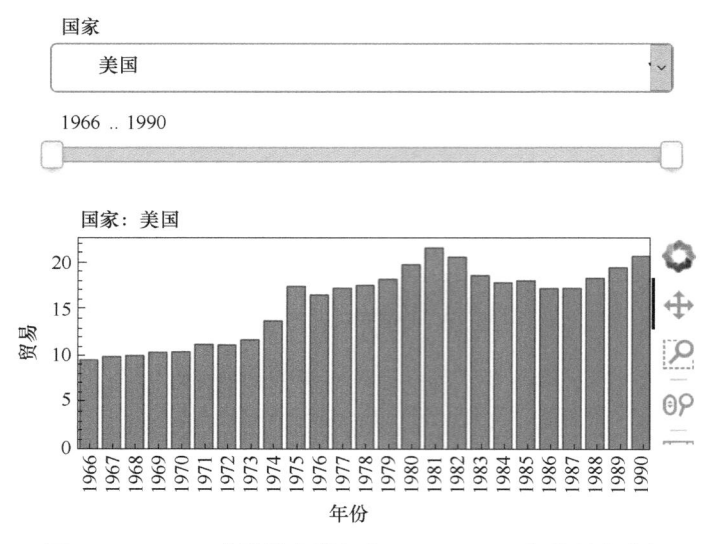

图 6.116　Panel 支持带有嵌入式 Holoviews 可视化的仪表板

6.6　Plotly

Plotly 是一个基于 Web 的可视化库，最方便使用的是 plotly_express。与普通的 plotly 相比，plotly_express 的主要优势在于创建常见图表更加简单方便，减少了繁琐的操作。

```
>>> import pandas as pd
>>> import plotly_express as px
>>> gapminder = px.data.gapminder()
>>> gapminder2007 = gapminder.query('year == 2007')
>>> gapminder2007.head()
      country continent  year  lifeExp       pop  gdpPercap iso_alpha  iso_num
11  Afghanistan     Asia  2007    43.83  31889923     974.58       AFG        4
23      Albania   Europe  2007    76.42   3600523    5937.03       ALB        8
35      Algeria   Africa  2007    72.30  33333216    6223.37       DZA       12
47       Angola   Africa  2007    42.73  12420476    4797.23       AGO       24
59    Argentina  Americas  2007    75.32  40301927   12779.38       ARG       32
```

用 scatter 函数绘制曲线图（见图 6.117）。

```
>>> fig=px.scatter(gapminder2007,
...                x='gdpPercap',
...                y='lifeExp',
...                width=400,height=400)
```

要绘制的 Dataframe 的列是使用 x 和 y 关键字参数进行选择的。width 和 height 关键字指定了图表的大小。fig 对象是一组 Plotly 指令，这些指令传递给浏览器，使用 Plotly JavaScript 函数进行渲染，其中包括一个常见图形功能的交互式工具栏，如缩放等。

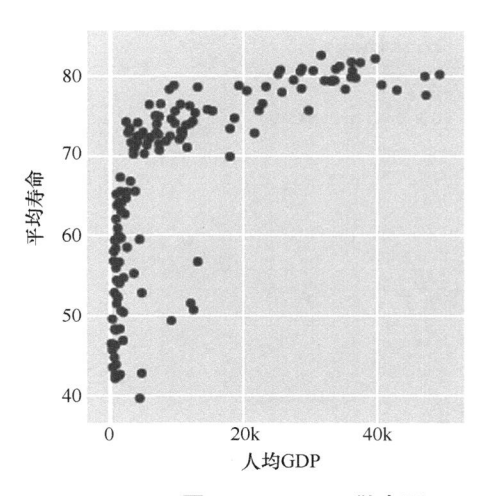

图 6.117　Plotly 散点图

与 Altair 一样，你可以给 Dataframe 列指定图形属性。例如将分类 continent 列指定为图中的颜色（见图 6.118）。

```
>>> fig=px.scatter(gapminder2007,
...                x='gdpPercap',
...                y='lifeExp',
...                color='continent',
...                width=900,height=400)
```

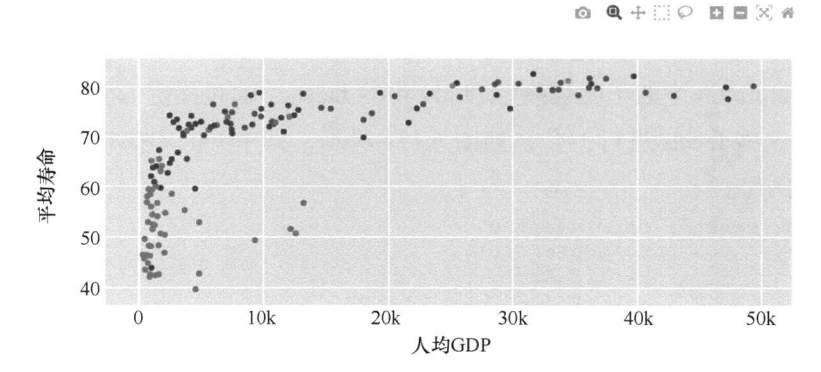

图 6.118　与图 6.117 相同的散点图，但现在各个国家分别着色

下面的代码中，使用 'pop' 列指定图形中标记的大小，并使用 width 和 height 关键字参数指定图形的大小，size 和 size_max 确保标记大小适合绘图窗口（见图 6.119）。

```
>>> fig=px.scatter(gapminder2007,
...                x='gdpPercap',
...                y='lifeExp',
...                color='continent',
...                size='pop',
...                size_max=60,
...                width=900,height=400)
```

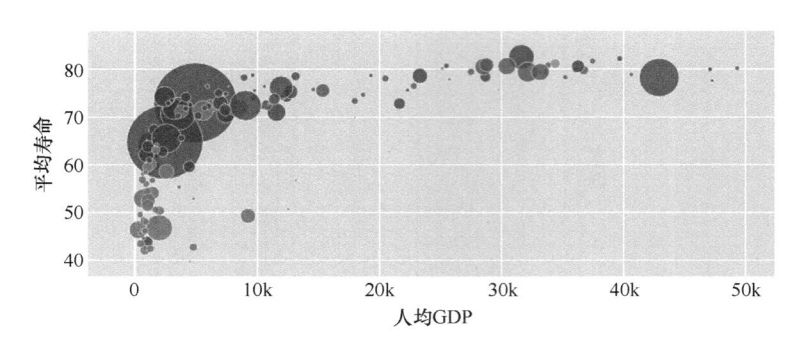

图 6.119　与图 6.118 相同，人口值缩放标记大小

在浏览器中，将鼠标悬停在数据标记上将触发带有国家名称的弹出式注释（见图 6.120）。

```
>>> fig=px.scatter(gapminder2007,
...                x='gdpPercap',
...                y='lifeExp',
...                color='continent',
...                size='pop',
...                size_max=60,
...                hover_name='country')
```

在 Matplotlib 中的子图被称为 Plotly 中的 facets。通过指定 facet_col，散点图可以根据这些 facets 进行拆分，如图 6.121 所示，其中 log_x 将水平比例尺更改为对数比例尺。

```
>>> fig=px.scatter(gapminder2007,
...                x='gdpPercap',
...                y='lifeExp',
...                color='continent',
...                size='pop',
...                size_max=60,
...                hover_name='country',
...                facet_col='continent',
...                log_x=True)
```

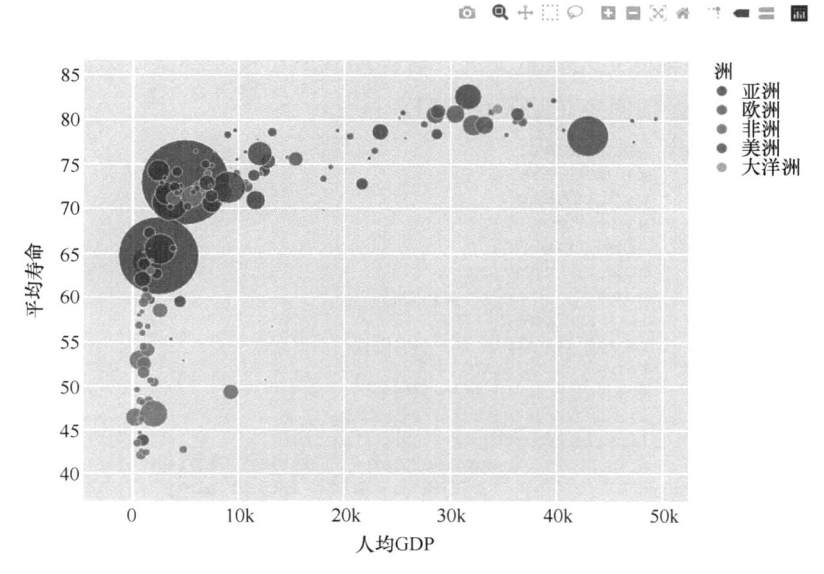

图 6.120　与图 6.119 相同，但标记上有悬停提示信息

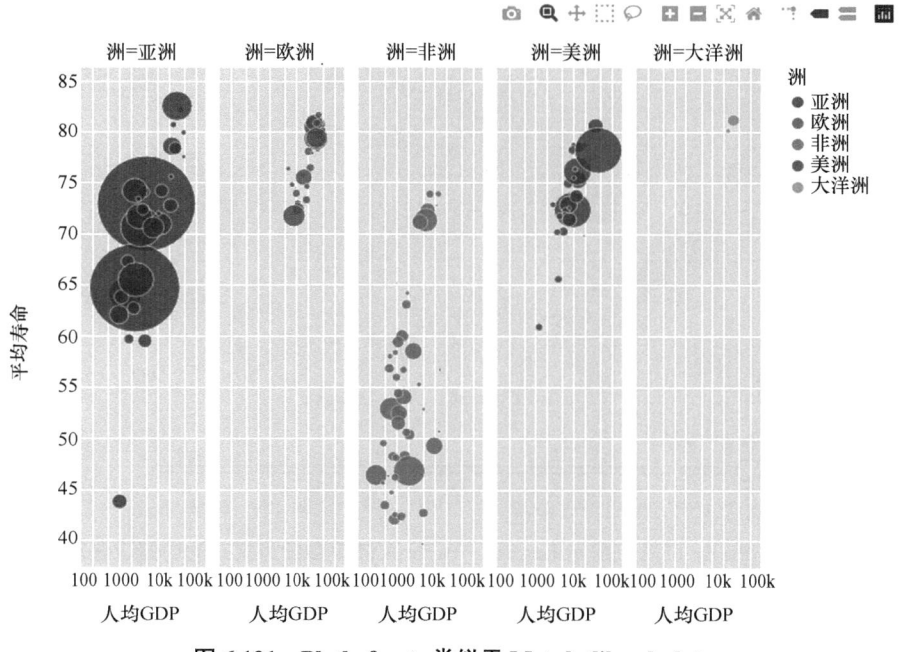

图 6.121　Plotly facets 类似于 Matplotlib subplots

Plotly 也可以构建动画。animation_frame 参数指示动画应该按照数据框中的年份列进行递增。animation_group 参数触发在每个动画帧重新呈现给定的组。在构建动画时，保持轴固定是很重要的，这样动画的运动才不会令人困惑。range_x 和 range_y 参数确保轴固定（见图 6.122）。

```
>>> fig=px.scatter(gapminder,
...                x='gdpPercap',
...                y='lifeExp',
...                size='pop',
...                size_max=60,
...                color='continent',
...                hover_name='country',
...                log_x=True,
...                range_x=[100,100_000],
...                range_y=[25,90],
...                animation_frame='year',
...                animation_group='country',
...                labels=dict(pop='Population',
...                            gdpPercap='GDP per Capita',
...                            lifeExp='Life Expectancy'))
```

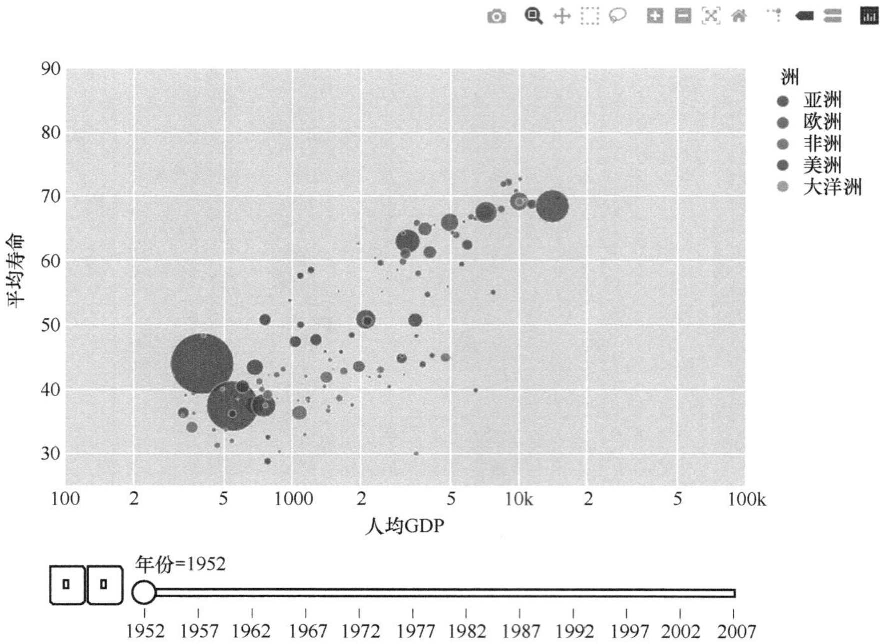

图 6.122　Plotly 可以使用 animation_frame 关键字参数为复杂的绘图设置动画

常用的统计图也很容易用 plotly_express 实现，示例如下：

```
>>> tips = px.data.tips()
>>> tips.head()
```

```
     total_bill   tip     sex smoker  day    time  size
0         16.99  1.01  Female     No  Sun  Dinner     2
1         10.34  1.66    Male     No  Sun  Dinner     3
2         21.01  3.50    Male     No  Sun  Dinner     3
3         23.68  3.31    Male     No  Sun  Dinner     2
4         24.59  3.61  Female     No  Sun  Dinner     4
```

按 smoker 划分的 tip 总和绘制直方图（见图 6.123）。

```
>>> fig=px.histogram(tips,
...                  x='total_bill',
...                  y='tip',
...                  histfunc='sum',
...                  color='smoker',
...                  width=400, height=300)
```

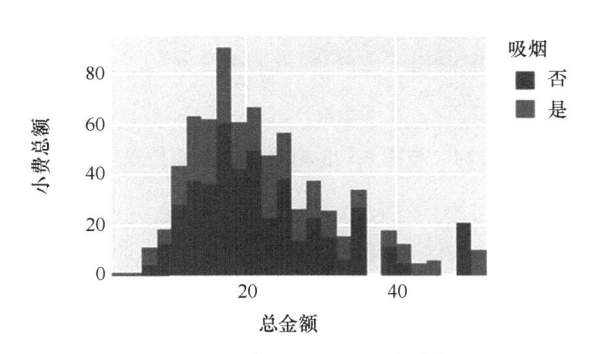

图 6.123 按是否吸烟分类的柱状图

Plotly 在浏览器中呈现图形所需的所有内容都包含在 fig 对象中；因此，要更改绘图中的任何内容，必须更改该对象中相应的项。例如，要更改图 6.123 中某个直方图的颜色，我们访问并更新 fig.data［0］.marker.color 对象属性。结果如图 6.124 所示。

```
>>> fig=px.histogram(tips,
...                  x='total_bill',
...                  y='tip',
...                  histfunc='sum',
...                  color='smoker',
...                  width=400, height=300)
>>> # 从变量中获取其他特征
>>> fig.data[0].marker.color='purple'
```

Plotly 可以绘制其他统计图，例如图 6.125 中显示的箱线图，使用 orientation 关键字参数将箱线图水平排列。

```
>>> fig=px.box(tips,
...            x='total_bill',
...            y='day',
```

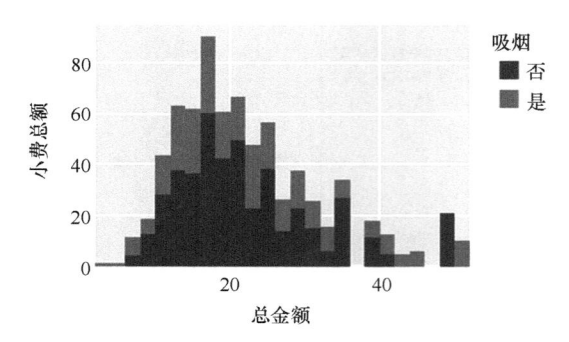

```
...              orientation='h',
...              color='smoker',
...              notched=True,
...              width=800, height=400,
...              category_orders={'day': ['Thur', 'Fri', 'Sat',
↪   'Sun']})
```

图 6.124　与图 6.123 相同，但有颜色变化

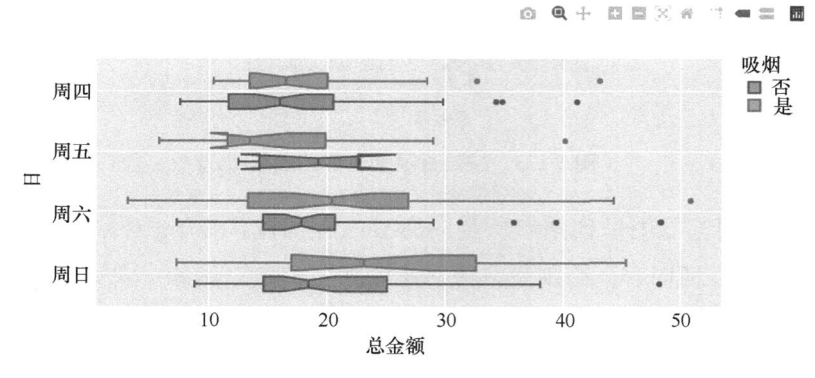

图 6.125　箱线图可以水平或垂直方向

小提琴图（violin plot）是另一种展示一维数据分布的方法（见图 6.126）。

```
>>> fig=px.violin(tips,
...               y='tip',
...               x='smoker',
...               color='sex',
...               box=True, points='all',
...               width=600,height=400)
```

边缘分布图（marginal plot）可以使用 marginal_x 和 marginal_y 关键字参数绘制，trendline 关键字参数表示使用普通最小二乘拟合来绘制中间子图中的趋势线（见图 6.127）。

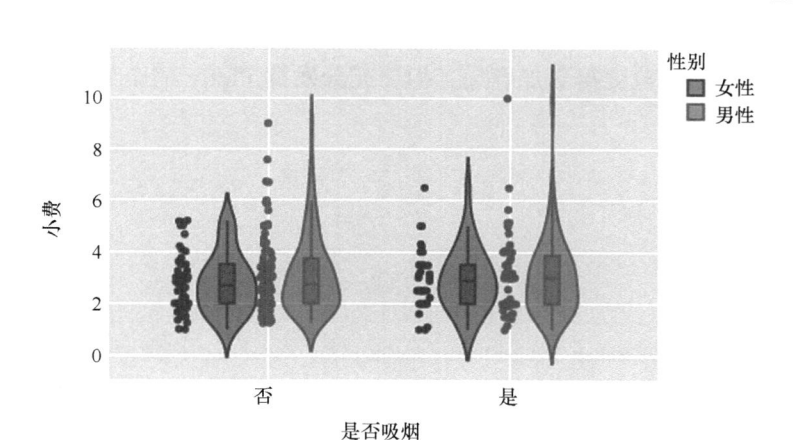

图 6.126　Plotly 支持一维概率密度函数可视化的小提琴图

```
>>> fig=px.scatter(tips,
...                x='total_bill',
...                y='tip',
...                color='smoker',
...                trendline='ols',
...                marginal_x='violin',
...                marginal_y='box',
...                width=600,height=700)
```

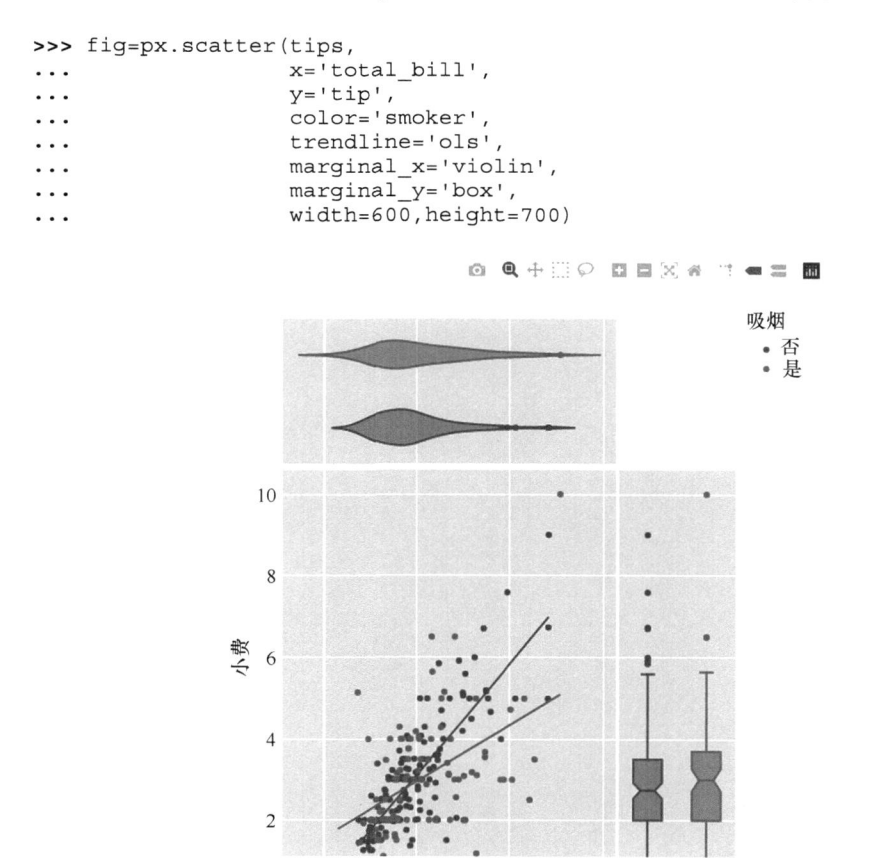

图 6.127　使用 marginal_x 和 marginal_y 关键字参数绘制边缘分布图

本章仅简单地介绍了 Plotly 的一些基本功能，如需了解更多新的图形类型可以通过查阅相关网站的支持文档。Plotly Express 使得生成由 Plotly JavaScript 库渲染的复杂 Plotly 规范变得更加容易，但要充分发挥 Plotly 的全部功能，需要使用裸 Plotly 命名空间。

参考文献

1. Bokeh Development Team. Bokeh: Python library for interactive visualization (2020)